National Climate Policy

W0234779

Failed attempts at producing ambitious global climate commitments and instruments have made it increasingly important for nation-states to deliver climate policies. This in turn requires a better understanding of national climate policymaking. In this book, Elin Lerum Boasson develops an innovative and well-grounded analytical framework for assessing national climate-policy development.

Why do national climate policies emerge and change? This question is underpinned by the role played by different actors and the kind of social mechanism at work. Boasson asks, to what extent and how is the emergence and change of climate policy influenced by: politicians and the national political fields; business and organizational fields; EU policy and the European environment; social and entrepreneurial mechanisms?

Combining policy studies with sociological new institutionalism, and drawing on three climate policy sub-areas in Norway: renewable energy, low-energy buildings and carbon capture and storage, Boasson presents a *multi-field framework* that allows the reader to capture the entire policy cycle, explaining policy initiation, policy adoption and the long-term, social feedback effects resulting from implementation (or lack of implementation).

Elin Lerum Boasson is a senior researcher at the CICERO Centre for International Climate and Environmental Research and the Fridtjof Nansen Institute in Oslo, Norway. Boasson has specialized in climate policy studies, combining assessment of national climate policymaking with insights into EU climate policymaking.

Routledge Research in Environmental Policy and Politics

1 **Green vs. Green**
The Political, Legal, and
Administrative Pitfalls Facing
Green Energy Production
*Ryan M. Yonk, Randy T. Simmons,
and Brian C. Steed*

2 **The Lilliputians of Environmental
Regulation**
The Perspective of State Regulators
*Michelle C. Pautz and
Sara R. Rinfret*

3 **Globalization, Political
Institutions and the Environment
in Developing Countries**
Gabriele Spilker

4 **Comparative Perspectives on
Environmental Policies and Issues**
Edited by Robert Dibie

5 **Framing Environmental Disaster**
Environmental Advocacy and the
Deepwater Horizon Oil Spill
Melissa K. Merry

6 **Reforming Law and Economy for
a Sustainable Earth**
Critical Thought for Turbulent Times
Paul Anderson

7 **National Climate Policy**
A Multi-field Approach
Elin Lerum Boasson

National Climate Policy
A Multi-field Approach

Elin Lerum Boasson

Routledge
Taylor & Francis Group

NEW YORK AND LONDON

First published 2015
by Routledge
711 Third Avenue, New York, NY 10017, USA

and by Routledge
2 Park Square, Milton Park, Abingdon, Oxfordshire OX14 4RN

First issued in paperback 2017

*Routledge is an imprint of the Taylor & Francis Group,
an informa business*

© 2015 Taylor & Francis

Library of Congress Cataloging-in-Publication Data
Boasson, Elin Lerum, 1978–
    National climate policy : a multi-field approach / by Elin Lerum Boasson.
        pages cm. — (Routledge research in environmental policy and
politics ; 7)
    Includes bibliographical references and index.
    1. Climatic changes—Political aspects.   2. Climatic changes—Social
aspects.   I. Title.
    QC903.B63 2015
    363.738'74561—dc23
    2014021157

ISBN 13: 978-1-138-05907-8 (pbk)
ISBN 13: 978-1-138-78113-9 (hbk)

Typeset in Sabon
by Apex CoVantage, LLC

Contents

List of Abbreviations   vii
List of Tables   ix
Preface   xi

1   A Multi-field Approach   1

PART I
Theory

2   Multi-field Social Mechanisms   25

3   Multi-field Entrepreneurship Mechanisms   62

PART II
Case Studies

4   The Power of Politics: Carbon Capture and Storage (CCS)   83

5   Entrepreneurship Paradoxes: Renewable Energy Policies   109

6   The Strength of a Pluralist Organizational Field:
    Energy Policy for Buildings   134

PART III
Comparisons and Final Conclusions

7   Comparative Assessment   161

8   Theory Conclusions   182

9 **Advice to Policymakers and Stakeholders** 199

Interviewees 207
References 213
Index 233

Abbreviations

BP	British Petroleum
CCS	Carbon Capture and Storage
CEN	European Committee for Standardization
CO_2	Carbon Dioxide
COP	Conference of the Parties (UNFCCC)
DG	Directorate (within the European Commission)
EFTA	European Free Trade Association
EEA	European Economic Area
EOR	Enhanced Oil Recovery
EPBD	Energy Performance of Buildings Directive
ESA	EFTA Surveillance Agency (ESA)
ETS	Emissions Trading System
EU	European Union
GHG	Greenhouse Gases
IEA	International Energy Agency
IPCC	Intergovernmental Panel on Climate Change
IPPC	Integrated Pollution Prevention and Control (EU Directive)
KRD	Kommunal- og reionaldepartementet (Norwegian Ministry of Local Government and Regional Development)
LO	Landsorganisasjonen (The Norwegian Confederation of Trade Unions)
MPE	Norwegian Ministry of Petroleum and Energy
NGO	Non-Governmental Organization
NVE	Norges Vassdrags- og energidirektorat (The Norwegian Water Resources and Energy Directorate)
OECD	Organization for Economic Co-operation and Development
R&D	Research and Development
SFT	Statens Forurensingstilsyn (The Norwegian Pollution Authority)
TCM	Technology Centre Mongstad

TwH	Terawatt Hours
UK	United Kingdom
US	United States of America
USA	United States of America
UNCCC	United Nations Framework Convention on Climate Change
ZERO	Zero Emission Resource Organisation

Tables

1.1	Typology of climate-policy measures	8
2.1	Four political field social mechanisms	39
2.2	Two political logics	41
2.3	Four organizational field social mechanisms	48
2.4	Professional logics related to climate-policy development	51
2.5	Four European environment social mechanisms	55
2.6	Comparing field and European environment characteristics	59
3.1	Core characteristics of the four entrepreneurial mechanisms	77
4.1	Mechanisms in the case of CCS	99
5.1	Mechanisms in the case of renewable electricity	124
5.2	Mechanisms in the case of renewable heating	124
6.1	Mechanisms in the case of energy policy for buildings	149
7.1	Climate-policy outcomes in the four Norwegian cases	162
7.2	Cross-case comparison of political fields	164
7.3	Cross-case comparison of organizational fields	167
7.4	Cross-case comparison of European environments	171
7.5	Entrepreneurial and social mechanisms in national political fields	174
7.6	Entrepreneurial and social mechanisms in national organizational fields	175
7.7	Entrepreneurial and social mechanisms relating to the European environment	178

Preface

This book marks the end of a long journey. I first started to explore how EU policy influenced Norwegian climate policies in 2006. The following year I received a PhD grant that allowed me to develop my initial thoughts into a full-blown thesis. The book you now hold is a considerably re-written version of the PhD thesis which I defended at the University of Oslo in December 2011.

Many people have contributed during my eight years of working on this manuscript. This study would simply not have been possible had it not been for the 95 interviewees. Their contributions have been of crucial importance to the project, and I am very grateful to all of you for your willingness to share your unique and valuable knowledge with me. Some of you spent several hours talking with me, and I have even interviewed several of you twice. You have given me insights that never could have been gained from written material alone.

A special thanks to my supervisor Tom Christensen, who has provided backing and masses of good advice throughout the process. Many other scholars have provided additional guidance and inspiration. Dick Scott, whom I had the pleasure of meeting in 2006, gave me the idea of the applying the neo-institutional term 'organizational field' actively in my research. Later, in 2009, he kindly agreed to provide input to some early chapter drafts. During my stay at the University California, Berkeley, from August 2009 to August 2010, I had the pleasure of participating in workshops led by Neil Fligstein at the CCOP (Center for Culture, Organization, and Politics). Neil also kindly gave me valuable input to early drafts of chapters. I am particularly grateful to him for the idea of exploring entrepreneurship in greater detail.

I have attended various conferences and workshops in which participants have given me valuable feedback. I would especially like to thank Sigrid Quack, Johan P. Olsen and Kendall Stiles. During the first five years in the life of this book project I was working at the Fridtjof Nansen Institute (FNI). Special thanks go to Per Ove Eikeland and Tor Håkon Inderberg. Thanks also to the Research Council of Norway and the industry partners (Energi Norge, Norsk Industry, Statkraft, Stanett and Svenska Kraftnät) that provided the funding for the larger CANES project that included my PhD.

I owe warm thanks to Trond Petersen, head of the Sociology Department at UC Berkeley, for inviting me to stay at the Norwegian Research Centre at Berkeley for a year and for providing exceptionally good working conditions. I arrived in August 2009 with two half-written chapters; one year later I had finalized the first draft of a full manuscript.

Arild Underdal, Sebastian Oberthür and Kerstin Sahlin did a superb job on my PhD supervising committee. All three contributed inputs that enabled me to sharpen my arguments and improve the manuscript. A special word of thanks to Arild, who also later as a colleague at CICERO, has provided additional guidance and help in finalizing the manuscript. I am also grateful to CICERO for giving me the opportunity to continue working on this book project, after I began work there in 2012.

In summer 2013, Miranda Schreurs kindly invited me, with family, for what was a highly enjoyable and productive one-month research stay at the Environmental Policy Research Centre, Freie Universität Berlin. Thanks also to Åsa Lerum and Anne Klock for taking care of our toddler Liv, so that her parents were free to do research.

In 2012 I joined the network on Climate Policy Innovation, led by Andy Jordan and Dave Huitema. Discussions in workshops arranged by this network have contributed significantly to improving my thinking, especially with respect to entrepreneurship. Fortunately, we have recently been awarded funding from COST (European Cooperation in the field of Science and Technology Research). I am convinced that the future collaboration in the 'Innovations in Climate Governance' (INOGOV, COST Action IS1309) action will enhance collaboration among researchers who study national climate policy development.

Three anonymous reviewers have provided helpful guidance on how to transform my thesis into a good book. I am specifically happy for the many pages of detailed comments from one of the reviewers; your suggestions proved very important to the final and last revision of the manuscript.

Susan Høivik has copy edited several versions of this manuscript. Thank you for your patience!

Lastly, let me thank my best friend, colleague, co-parent and partner, Jørgen Wettestad. Not only do you make everyday life good, fun and interesting, Jørgen—you also provide crucial input to my academic work. Thank you for respecting my way of thinking, for remaining highly critical—and for cheering me up when I most needed it. I am particularly grateful for the many good comments and advice you gave during our wonderful research stay in Berlin, where I re-wrote the whole book.

All the people mentioned in this preface have helped to make this book possible. I am happy that none of you told me in advance that it would take eight years. Needless to say, any remaining flaws in the resultant book are my own responsibility.

Oslo, Blindern, 23 May 2014

1 A Multi-field Approach

1.1 INTRODUCTION

National climate policymaking varies from issue-area to issue-area—renewable energy, energy efficiency and carbon regulation are embedded in different national climate-policy 'landscapes' with varying characteristics. One may be alpine, with steep valleys and towering peaks, glaciers and rivers *without* bridges. Others are formed as rolling mountain plateaus, where a hiker can walk for hours without encountering other human beings. Or there may be deep, trackless wilderness forests. The skills and gear needed to master climate-policy development will differ from landscape to landscape.

This book presents a *multi-field framework* that can help policymakers and analysts to map and understand the kind of climate-policy landscape they are faced with, and the skills and gear required. Of course, national climate policy is not literally made up of hard rock, river streams and wildlife. The materials are slightly more malleable: human beings, institutions and organizations. The multi-field framework presented here provides a foundation for systematic, comparative and cumulative research on climate policy, as well as giving policymakers tools they can use to understand, navigate and shape the processes of national policy development.

Written primarily within the tradition of political science, this book also draws on insights developed in the field of sociology, particularly neo-institutional sociology. The multi-field framework profits from two pervasive trends in social science: first, many scholars have increasingly emphasized the role of *fields* (or 'segments', 'policy systems' or 'policy monopolies'), showing how such fields distribute power among actors and shape values, identities and interests. A societal field is seen as a circumscribed sphere of political and social life with an identifiable social architecture, a distinct social order, a particular constellation of actors, identifiable distribution of resources and discrete cultural features, influencing which climate-policy actions people perceive as rational and appropriate. Climate policymaking will be influenced by multiple fields, with the number of fields involved and their characteristics varying between policy areas.

Second, sociologists and political scientists alike have begun to pay greater attention to *entrepreneurship*—which they may also call 'networking', 'advocacy', 'agency', 'persuasion' or 'institutional work'. By skilfully using entrepreneurial techniques, actors can modify the influence of the involved fields and may gain greater influence on policy development.

Mitigation of climate change is a central political issue of the 21st century and an important academic challenge. Natural scientists have provided us with a good understanding of the nature of climate change, and engineers and economists have proposed many technical solutions and policy instruments. However, political scientists and sociologists have to a very limited extent informed policymakers on how to develop political processes and solutions for dealing with climate change. This is highly regrettable. De-carbonizing the economy will require deep-going transformations in our energy systems, our industries and our political-administrative systems. The social sciences possess insights that may help policymakers to facilitate this major transition process.

Climate-policy discussions today are often characterized by blame games, with each side accusing the other of lacking willingness to solve the problem. The multi-field framework goes beyond these over-simplified lines of conflict, recognizing that national climate policy is very seldom a fight between 'the good' and 'the bad'. Complex social structures tend to constrain and enable developments in national climate policy. Yes, in many countries, denial of climate change still features in policymaking—but lack of acceptance or willingness to cope with climate change is rarely the main reason why governments fail to develop ambitious, stable national policies. Rather, disagreements on how such policies should be designed often obstruct both the adoption and the stability of climate policies.

The multi-field framework recognizes that national policymaking processes are complex and cannot be explained by neat, precise and parsimonious correlations between explanatory factors and outcomes. It is a mechanism-based approach, which means it aims to underpin generalizations about *the process* of national climate-policy development (see Hall 2003:393; George and Bennett 2005:140). Mechanism-based explanations focus on observed regularities, seeking to render them intelligible by specifying the relationship between the explanatory factors and outcomes (see Hedström 2008). The aim is to develop mechanism-based explanations that are abstract as well as realistic, but do not take methodological individualism as the starting point (see Abbott 2007; Bhaskar 1998; Gross 2009:384).

The overarching analytical question of this book is: *Why do national climate policies emerge and change?* This question is underpinned by four sub-questions concerning the role played by different kinds of actors and the kind social mechanism at work. We ask: to what extent and how is the emergence and change of climate policy influenced by

1. politicians and the national political fields?
2. business and organizational fields?
3. EU policy and the European environment?
4. social and entrepreneurial mechanisms?

The first sub-question concerns the political steering and political control over a climate policy. A political field comprises legislative assemblies and committees, political parties, governmental executives and the political leadership of relevant government ministries. The second question will lead to examination of the importance of industry and business actors embedded in organizational fields. Government agencies, regulators, industry, business organizations and NGOs are all involved in organizational fields. The third sub-question deals with how national policymaking in Europe is influenced by the broader constellation of political and organizational fields in the 'European environment'. This 'European environment' is a looser term, encompassing the national political and organizational fields in states that are members of, or otherwise affiliated to, the European Union,[1] EU policy development arenas and relevant European organizations.

Fourth, this book will explore the relative importance of slow-moving social processes and agency. Special intention will be given to *entrepreneurship:* acts aimed at enhancing policy influence by altering the distribution of authority and information, and/or aimed at altering norms and cognitive frameworks, worldviews or institutional logics.

Answers to these four questions are important not only to climate-policy studies, but also to the wider political science debate on the relative power of politicians, business and international features in national policy development processes.

In this book, the multi-field framework is applied in a comparative case study of four central climate policies—carbon capture and storage (CCS), renewable electricity, renewable heating and energy policy for buildings—and how they unfolded in Norway from 2000 to 2010. These specific cases were selected because they differ with respect to several points of interest, including the power of politicians, the role played by organizational fields and the influence of the European environment.

Let us now take a look at the state of the art in national climate-policy studies, before turning to the main components of the multi-field framework, and the empirical case studies.

1.2 NATIONAL CLIMATE POLICY

State of the Art

Climate change is a very serious matter indeed. Why, then, has this challenge to a large extent been met with non-action? Sociologists and psychologists have examined why we seem to be incapable of acting on the overwhelming scientific information now available, and political scientists have explored how some organized actors, like the petroleum industry and climate-change deniers, have obstructed developments in climate policy (see Giddens 2009;

Norgaard 2011). Such studies are important, but they disregard the substantial climate policies that actually have been adopted and implemented. Rather than adding to the non-action literatures, in this book I present a foundation for assessing the national political processes that have yielded tangible results.

The most extensive literature on climate politics centres on the international climate regime. This literature has repeatedly argued that it will be extremely challenging, if not impossible, to develop ambitious international policies to mitigate climate change (see e.g. Delmas and Young 2009; Underdal 2002). That argument rests on several assumptions widespread in political science—not least, that all actors will seek to defend their short-term economic interests, and that national politicians act as mere instruments of dominant national industries (see discussions in Hooghe and Marks 2001; Moravcsik 1998; Skocpol 1985). International climate politics have been characterized by the search for one miraculous universal solution—unsurprisingly, with little success.

Global climate regime studies are part of the larger literature on international environmental regimes. After an initial focus on single case studies, there has come greater use of cross-case work within the literature on environmental regimes—for instance, comparing the global climate-policy regime with other international environmental agreements (see Miles et al. 2002; Stokke 2012). We can also note the emergence of a comparative literature on EU climate policy (see Boasson and Wettestad 2013).

Such comparative research strategies have not been as prevalent in the literature on national climate policy, which has been characterized by qualitative and empirically rigorous case studies. Many single case studies have been conducted, but they hardly speak to each other, and no particular explanations dominate the field of study. Research teams, such as Townshend and colleagues (2013), have mapped the climate policies of larger numbers of countries, but few offer a conceptualization of possible explanations of the similarities and differences in climate-policy outcomes (see Dubash et al. 2013 for an overview).

The UN International Panel on Climate Change (IPCC), Working Group III Report on National and Sub-national Policies and Institutions, summarizes some of the qualitative contributions and the more recent, but fewer, quantitative assessments of national climate policies (Somanathan et al. 2014). This report documents steep growth in national climate policies after 2007 (see also Dubash et al. 2013) and explains the difficulties related to measuring the effectiveness and efficiency of the various measures. Further, it shows that even though carbon pricing has been widely discussed in the literature, sector-specific measures still prevail. Most typically, climate polices are introduced 'through a sector process led by relevant government departments' (Somanathan et al. 2014:11).

The IPPC report contains a brief discussion of factors driving national climate-policy development, indicating that these factors include 'international pressure, scope for co-benefits and changing norms and ideas' (Somanathan et al. 2014:17). The IPCC authors draw heavily on a study by Erick Lachapelle and Matthew Paterson (2013). Lachapelle and Paterson compare the climate policies of large emitters and conclude that 'a good number of policy types (e.g. regulations and standards, economic incentives and, to a lesser extent, R&D) are relatively indiscriminate with regard to the various comparative features of states'. They show how factors with high explanatory value in other comparative politics literatures are of less value in comparative climate-policy research.

Peter Christoff and Robin Eckersley (2011) have conducted a comparative exercise with a dataset of 20 countries, quantifying and comparing climate-policy outcomes and briefly discussing some possible explanatory factors common to the comparative politics literature, such as regime type, national political system, national interests and national discourses. They conclude:

> [T]he quest to find a single cause, or even a common set of drivers, to explain climate leaders or laggards is a near-futile exercise (. . .) While there is no clear institutional 'identikit' of a likely climate leader, some general conditions for strong climate performance are discernible. These usually include an advanced economy, a strong civil society, a strong and respected tradition of scientific research, and a diverse media.
>
> (Christoff and Eckersley 2011: 444–45)

However, the explanatory factors highlighted by Christoff and Eckersley have limited value, as almost all OECD countries will fit with their descriptions. That said, the literature on national climate policy is more extensive when we take the rich case-study tradition into account. A great many single case studies of the climate policies of various countries have been published in national political science journals and environmental policy journals. Researchers such as Bang (2010), Compston and Bailey (2012), Harris (2007a), Harrison and Sundstrom (2010), Rabe (2008), Selin and VanDeveer (2009), and Wurzel and Connelly (2011) have studied national-level climate policy in Europe, as well as national and state-level climate policy in the USA. Most of these contributions are explicitly comparative, but without systematic assessment of the national-level policy dynamics steered by or centrally involving states. A few book-length studies have compared the climate policies of major climate-policy actors, with a specific focus on Japan, the USA and the EU (see Harrison and Sundstrom 2010; Schreurs 2002; Watanabe 2011).

These studies use widely diverging explanatory frameworks and the authors tend to pay more attention to describing specific national circumstances than to developing general explanatory frameworks. Without doubt, these studies have significantly enhanced our understanding of specific, real-world policy changes, showing how many different explanatory factors influence the development of climate policy. Several authors underscore that national policy outcomes are almost always the result of crossover effects between domestic and international politics (see Harris 2007b:31), and preference formation is recognized as pivotal. Harris (2007c:403) concludes that 'ideational variables are often very important or even central' to understanding national climate-policy outcomes in Europe. Similarly, Christoff and Eckersley (2011:442) argue that that issue framing, particularly in the early phase of the policy cycle, plays a major role in shaping the subsequent path of policy development.

State-led, national government is not the sole locus for innovations and decision-making on climate policy (see Bulkley and Moser 2007; Aall, Groven and Lindseth 2007; Hoffmann 2011). However, state-led action will have to be central if Greenhouse Gas (GHG) Emissions are to be cut to the extent necessary for keeping the rise in global temperature under 2°C (see Solomon et al. 2007; Stocker et al. 2014). For a more comprehensive collective understanding of climate governance, it is essential to take account of the climate-mitigation actions pursued by nation-states (Jordan and Huitema 2014). The authors of the IPCC report on climate policy conclude that 'institutional and governance changes can accelerate a transition to a low-carbon paths' (Somanathan et al. 2014:5) Moreover, it has become increasingly clear that 'a focus on the national level is a necessary first step to begin to flesh out the new geopolitics of climate change' (Lachapelle and Paterson 2013:567).

The political science literature on climate-policy development is still in its infancy. We know that many actors and factors may influence national policy, that initial framing is central to creation of interests, and that international factors play into the national processes. Less is known about the relative importance of national politicians, administrators and civil servants, industry and international factors. Climate-policy research needs improved, more elaborate theory frameworks: in order to be able to generalize about national climate policymaking at all, we must first identify and specify the mechanisms that shape the development of policy (Bennett 2008:704).

This book develops an analytical framework that may foster more systematic cross-country and cross-sector climate-policy assessments. The multi-field framework offers a conceptualization of how policy outcomes can assume a range of characteristics and may vary along several dimensions. Further, it shows how different issues of climate policy tend to be embedded in different multi-field clusters, or climate-policy landscapes. It also helps to distinguish between how strong social trends and how entrepreneurship may influence climate-policy development. The multi-field framework can help us to recognize the complexity of climate policymaking, understanding how preferences form and change over time, and why some actors may have the upper hand in some climate policies but fail to influence others.

Capturing Climate-Policy Variation

Cross-country and cross-sector differences in climate-policy measures have received scant attention in comparative climate politics literature (see Lachapelle and Paterson 2013:548). Neither the national nor the international literature takes the discussion of climate-policy measures as the starting point: these assessments tend to have 'the level of ambition' as their explanatory focus (see Underdal 2002; Wurzel and Connelly 2011; Christoff and Eckersley 2011).

The level of ambition is often determined by comparing policy outcomes with scientific advice, but persistent disagreement on what should be seen as acceptable levels of carbon emissions from specific countries has made this intrinsically challenging. Further, given the overwhelming character of the climate challenge, most climate policies seem far from adequately ambitious for putting the scientific advice into practice. Researchers often end up explaining why policy ambitions have been so low, and why certain policy recipes failed to be adopted. Scientifically explaining non-events is notoriously difficult—it is actual policy decisions and events that are amenable to research. However, because different sectors and different countries face different challenges in achieving low-carbon societies, it is demanding to develop a standard for level of ambition that can cover more than one country or even more than one sector. In order to create a foundation for comparative research, we need to focus on other, less normative dimensions of policy outcomes. What I propose here is to ground national policy studies in a classification of various climate-policy outcomes, drawing on insights from the environmental instrument literature.

Debate over the theoretical advantages and disadvantages of specific instruments has dominated the literature on climate and environmental policies (see discussions in Hood 2007; Jordan and Schout 2006). There is a vast literature on economically optimal climate-policy measures. Political scientists have developed complex policy typologies, giving weight to market measures as well as traditional regulation and information measures (Jordan et al. 2011). The IPCC report on national policies presents a typology which is quite representative: economic instruments, regulation and standards, information policies, government provision of public goods and voluntary actions (Somanathan et al. 2014:17–19). When the aim is to categorize measures into different policy areas or countries, it is advantageous to have fewer categories and a more systematic presentation of similarities and differences between categories.

I argue that most national climate-policy outcomes can be characterized along two central dimensions of state steering: technological vs. economic steering, and indirect vs. direct state steering. By combining the two categories, we get a typology of four climate-policy outcomes that captures a broad variety of policy outcomes. We begin with the dimension concerning the directness of state steering: whether the government should steer directly or indirectly (see Salamon 2002). On the lowest end of state engagement we find policies that leave the actual choices and specific decisions on practice and the use of technology to non-state actors. Such policies provide fairly general technical or economic steering signals, so it is up to private actors to translate these into action. At the opposite, high, end of state engagement are policies that strictly require certain practices or technologies—or the state itself may engage in, say, carbon mitigation, through state owned corporations or involvement in joint ventures with private actors.

The second dimension concerns the 'condensed form of knowledge about social control and ways of exercising it' on which policy outcomes are based (Lascoumes and Le Gales 2007). The technical-economic aspect here has two extremes. At the technical end, climate measures are designed to diffuse specific technologies or practices, and the policy relies on technical specifications (see discussion in Metz et al. 2007). The policy will set technical performance standards, whereas the societal costs involved will result from the technical requirements. And at the economic end, governmental regulations create or regulate economic incentives, whereas the actual technologies and techniques applied will be those that lie below certain economic thresholds.

Combinations of the two dimensions create the four categories presented in Table 1.1: *technology standards, market measures, governmental industry development* and *fiscal incentives*.

Technology standards are policies that give clear guidance but leave the more detailed decisions to private actors. This category includes regulations and standards that give private actors guidance on which practices, technologies or emissions levels are deemed acceptable. Pollution permits is a classic example of this kind of measure: it gives a cap on emissions, while leaving it to the private actor to decide which technology or practice to apply in

Table 1.1 Typology of climate-policy measures

Policy criteria ⇨ State Steering ⇩	Technological	Economic
Indirect state steering	**TECHNOLOGY STANDARDS** Governmental regulation of industry performance and practice.	**MARKET MEASURES** Emission trading schemes and certificate schemes.
Direct state steering	**GOVERNMENTAL INDUSTRY DEVELOPMENT** Ban and requirements of certain practices and technologies. Governmental companies and agencies directly involved in low-carbon industry transition. Feed-in support.	**FISCAL INCENTIVES** State aid schemes based on economic, not technical, criteria.

order to keep emissions below a certain threshold. Many land-use planning regulations also belong in this category. The technology standards category includes some measures that are often categorized as voluntary or information measures in the environmental policy literature. For instance, agreements between state and industry that create a cap on emissions will fit here. Voluntary agreements of a very loose nature, without any clear guidance on practice and technology, are not regarded as policy measures, however, and do not fit in this category. Information campaigns that give consumers technical and practical advice on specific issues, for instance on home insulation, are regarded as under 'technology standard'. General information campaigns, for instance aimed at promoting awareness of climate change, are not specific measures and do not fit here.

Market measures are governmental regulations aimed at creating economic incentives that can stimulate shifts in practices and technologies so as ultimately to reduce greenhouse gas (GHG) emissions. These measures create new markets or change the functioning of existing markets. This definition is broader than the traditional definitions of market instruments applied in environmental economics. For instance, the OECD defines market instruments thus:

> Market-based instruments seek to address the market failure of 'environmental externalities' either by incorporating the external cost of production or consumption activities through taxes or charges on processes or products, or by creating property rights and facilitating the establishment of a proxy market for the use of environmental services.
>
> (OECD 2007)

Both the classic understanding of market instruments and the definition of market measures presented above include energy and carbon taxation, emissions trading and market-based support schemes, such as green certificates: these measures leave it to private actors to select and develop the technologies and practices that are most profitable, given the new price signals introduced by government measures. With respect to emissions trading and green certificate schemes, the government does not set the economic incentives: these are set by market forces. In addition, governments may influence markets by launching product-labelling schemes (like energy labelling of electric appliances and buildings) and information campaigns (targeting a particular kind of producers and customers). The latter measures are not normally taken into account by environmental economists. Because they are designed with the objective of influencing behaviour through creating of market changes, they fit in this broader understanding of market measures.

With *governmental industry development,* the government engages specifically in industrial decisions concerning the choice of technology and practices. The government first determines that a certain technology or industry practice is to have priority, and then regulates technology choices directly, rather than regulating the acceptable costs of the various choices of technology. This may occur through strict regulations, where failure to meet the technological requirements will trigger some form of punitive action. Alternatively, state-owned companies or agencies may be directly involved in the development of new technologies and practices. Traditional feed-in schemes for renewable energy will also fall under this category: here the government interferes directly by setting the price on specific technologies (Commission 2005a, 2008a).

Fiscal incentives refer to state aid schemes where support is granted on the basis of economic, not technical, criteria. In such schemes, the government determines the economic criteria for selecting the projects eligible for support, and the state then undertakes the actual selection. Such support will be granted for the least costly, or most cost-efficient, renewable energy projects, energy efficiency measures or direct emission-reduction efforts.

This fourfold typology gives us a basis for categorizing policy outcomes in different policy areas. Whereas some policy areas are dominated by measures that fit within one of the categories, other areas may involve a combination of different policy categories. Further, the climate policies of some countries may be dominated by one type of measures, whereas we may see more variation elsewhere.

This typology can also be used to track policy change over time. For instance, the renewable energy policy of a country may fit the same category for a long period of time, whereas there may be shifts from one category to another in another country. Some climate policies prosper along the initial path, some encounter political deadlock before they are even realized, and yet others may be criticized so heavily that they get revised and changed. Here we are particularly interested in changes in the very existence and character of the policies, not in changes in the underlying normative ambitions. For instance, the character of a given state aid scheme may remain stable even though the government channels more (or less) funding into it. Some countries have changed their climate policies dramatically over time (for instance the Netherlands and UK); others have had more stable climate policies that have developed incrementally (as with Germany) (see Boasson 2013; Harris 2007a; Wurzel and Connelly 2011). Moreover, a country may have unstable climate policies in some sectors, and stable ones in others.

James Mahoney and Kathleen Thelen (2010) have shown how policy stability may also vary significantly from issue-area to issue-area. In some areas we may see straightforward *displacement*: the removal of one policy measure and the introduction of new rules that fit with another policy category. For instance, renewable energy policy may shift from a feed-in scheme to a green certificate scheme. Alternatively, new rules may be introduced on top of or alongside existing ones: *layering* of measures belonging to several categories. This will be the situation if two or more of the climate-policy measures presented in Table 1.1 are applied within the same issue-area. For instance, many countries combine various measures regarding energy policy for buildings with market measures on the top of technology standards (see Boasson 2013). Moreover, we may see *conversion:* changed enactment of existing measures due to their strategic redeployment. Rules developed for purposes unrelated to climate policy may gain significance in climate policy, or rules that originally fitted one of the climate-policy categories get deployed in ways that change the character of the policy. For instance, regulations originally created to deal with energy security or other environmental issues may be used to achieve climate-policy objectives.[2]

The Multi-field Policymaking Process

Climate-policy 'landscapes' vary significantly across issue-areas, countries and time. Each landscape is made up of multiple fields, with distinct societal orders and actor constellations; and specific distributions of structural resources and institutional and cultural features are involved in the development of national policy (Fligstein 2008; Scott [1995] 2001). The multi-field framework has been inspired by field theories in sociology, policy studies and by political science multi-level governance studies. It combines insights from various political science traditions and draws heavily on the neo-institutional sociological literature. In this book I use the terms 'multi-field' and 'policy landscapes' interchangeably: both terms refer to the same phenomena.

First, the framework focuses on the political field and how it contributes to shaping policymaking over time. To date, there has been considerable political science attention to elections and democratization, but less as regards the actual policy-shaping power of politicians; and policy studies have dealt more with the influence of societal actors than with politicians and their influence. Because politicians are democratically elected—and are expected to work at solving various central moral, industrial and economic dilemmas of the time—it is important to explore their independent effect on national policymaking (see Olsen 2010). Rather than assuming that politicians act as the instruments of powerful societal actors, or persuasive lobbyists, the multi-field framework specifies how and to what extent politicians may influence policy development under different conditions: when can we expect political actors to have a significant independent effect on policy development? And when will they merely respond to developments in other fields?

Politicians may accomplish impressive feats—but, I hold, their achievements will depend just as much on the opportunities available to them as on their personal ability and motivation to influence policy. The multi-field framework helps to explain how politicians tend to approach different climate-policy issues, and when they will generally have the upper hand. Moreover, it sheds light on why politicians are more willing to spend time on certain issues of climate policy than on others.

Second, political scientists have attached considerable weight to how industry, societal groups and government agencies contribute in shaping national policy outcomes—but disagreement runs deep. Some see society as composed of stable and powerful segments (or 'iron triangles' or policy monopolies) that allow industry actors to capture national policy developments; others describe industry and government as independent of each other, all having time and resources to compete for political influence. Moreover, some argue that business will always have a privileged position in policymaking. In contrast, others hold that various societal actors may influence policy development and that actors are tied to each other through loose networks. The multi-field approach makes possible a nuanced discussion about how, through which mechanisms business, civil servants and environmental organizations influence policy emergence and policy change.

With this book I develop a more refined conceptualization of how industry and government affect national policymaking. I introduce the neo-institutional sociology term 'organizational fields'—here understood as issue- or industry-specific configurations of governmental and private organizations, marked by a certain structural interrelationship and certain shared institutional understandings. This term takes into account the fact that industry and governmental organizations have a symbiotic relationship to each other: both groups of actors evolve and function in accordance with governmental regulations, and they may develop shared worldviews and preferences.

Further, the multi-field framework specifies how and to what degree organizational fields with different structural and institutional configurations—and hence different social orders—contribute to shaping policy outcomes. Such insights can help us to understand when, and under what conditions, business on the one hand and civil servants on the other will have a major say in policymaking. Particular attention will be paid to examining the conditions under which business will influence policy development the most. Moreover, this framework helps to explain why some areas of climate policy develop without much further ado, whereas others may be characterized by deep conflicts.

Third, the multi-field framework provides tools for assessing how international policies and other external factors may influence and alter the distribution of power in national policy development. Here, particular attention will devoted to the European environment—seen as an aggregate of several fields, encompassing all national organizational and political fields that relate to a specific issue-area and the fields related to Brussels policymaking. Whereas many studies have focused on variance in national compliance with EU regulations (see Mastenbroek 2005), far less is known about the actual role of Europeanization in national policy processes. In 2006, Börzell and Risse concluded: '[w]e hardly know anything about how the emergence of a European structure of political and societal interest representation impacts on the processes of political contestation and interest aggregation in the member states' (2006:488). The ensuing years have not brought major improvements.

This book supports a broad approach to exploring how the European environment plays into national policymaking. Issue-specific European environments may differ in their institutional features, as well as in the distribution of their structural resources. In practice, all the national fields relating to an issue-area will take part in the European environment.

The multi-field framework can help to fill in the gaps in our understanding of the mutual interrelationship between national-level and European-level policy development, and how the national fields and the European setting may interact over time. Moreover, it shows how, in some exceptional situations, the EU may have the upper hand in climate policymaking and neither organizational nor field-level national actors have much leeway to alter the development of policy—though national actors will normally have the upper hand in the development of national climate policy. That said, entrepreneurs may deliberately interpret European developments in line with their own preferences, as an effective way of gaining greater political clout.

The multi-field framework is intended primarily for assessing national climate policies in countries that are members of the EU, or otherwise affiliated to it. It could be argued that the situation in Europe today is very special due to the *sui generis* (unique) character of the EU and the political order, and that the multi-field framework is therefore applicable only in Europe. However, whether the EU is unique in its characteristics or whether it can be compared to either other international organizations or federal governments (such as in the USA) is contested. I hold that many of the concepts, typologies and propositions presented in this book are likely to have relevance to other geographical areas as well. It is of considerable interest to explore whether the national climate policies of countries outside the EU and the European Economic Area (EEA) are influenced by international factors in similar or different ways to what we see in Europe.

Social and Entrepreneurial Mechanisms

Some actors are better than others at navigating the climate-policy landscapes and influencing policy outcomes. Many enthusiastic and skilled actors with bright ideas have delved into climate-policy processes, and some have been successful. But quite a few have ended up in apathy, fatigue and lack of belief in the ability of the political-administrative apparatus to solve the many problems linked to climate change. Several people I interviewed during the writing of this book were of the latter type: indeed, some begged me to find out why they had failed and to help them understand the actions of other stakeholders and policymakers. It is evident that strong historical trends, resource distribution and entrenched cultural patterns will all influence and constrain actors seeking to influence the development of climate policy. Yet, climate-policy landscapes are more malleable than natural landscapes. They are governed by social mechanisms which, unlike the laws of physics, can be influenced and changed. But—when is entrepreneurship crucial for policy development, and when not?

My approach specifies the relative importance of social and entrepreneurial mechanisms, recognizing that, whereas social mechanisms are generally fundamental to policy development, entrepreneurs may occasionally intervene and shift the direction of policy development. The social sciences today are rife with disagreement as to whether the main drivers of social change are societal factors or individual, strategic actions. Some lines of study have seen the historically developed distribution of structural resources and institutional, cultural patterns as main drivers; others see skilled social actors as the architects of political and institutional change. In the multi-level governance approach—as also in the network and agenda-setting approaches to national policymaking—considerable weight is accorded to entrepreneurship: individual skills like agenda-setting and lobbying abilities. In contrast, many studies of Europeanization and segmentation approaches to policymaking have focused on social mechanisms, like the distribution of finances, information, authority, ideas and norms.

My general assumption is that the social influence of multiple fields and entrepreneurship represents two fundamentally different, but related, policy-shaping mechanisms. Rather than trying to single out which one is more important, multi-field framework underpins explorations of the relative importance of outstanding individual achievements *and* social mechanisms. The position of 'entrepreneur' is not seen primarily as a disposition or a quality of an individual: it is a role that becomes available under certain conditions (see Fligstein and McAdam 2012:181).

Good mechanism-based explanations should be both abstract and realistic, clarifying specific political phenomena on the basis of explicitly formulated theories of action and interaction (see Hedström 2008:333). Mechanisms govern the process of policy development: they are the paths or features that connect causes and effects (Hedström 2008:320, see also Danermark et al. 2002:55; George and Bennett 2005:137; Hall 2003). Mechanisms are not readily evident through empirical observation: they need to be conceptualized and refined by the researcher (Hall 2003). True, mechanism-based explanations will always be imprecise representations of the processes that unfold in real life, so they will always be open to improvement. That, however, should inspire us to constant revision and re-thinking. The only way to advance beyond the boundaries of our current knowledge is to make our assumptions and conceptualizations as accurate and explicit as possible, and then work constantly to improve them (George and Bennett 2005:142).

I operate with an analytical distinction between *entrepreneurial* and *social* explanations. Social mechanisms are understood as slow-moving processes, effectuated by actors who follow the institutional patterns of their organizations and fields and act on the basis of their structural positions. Entrepreneurship—referring to acts aimed at enhancing policy influence by altering distribution of authority and information, and/or activities aimed at altering norms and cognitive frameworks, worldviews or institutional logics—may be performed by individual persons as well as organizations or sub-organizations. Assessments of climate policy should investigate both entrepreneurial and social mechanisms, recognizing that the two types of mechanisms are interrelated.

This book specifies the social conditions under which entrepreneurial possibilities may unfold, and when entrepreneurship is likely to have the greatest influence on policy development. Politicians seldom have the time and energy to act as entrepreneurs, but when a climate issue is seething with political competition, politicians may decide to become involved—and may demonstrate extraordinary entrepreneurial strength and creativity. More often, field-level actors will engage as entrepreneurs. The European environment will produce a range of entrepreneurial opportunities, and those who know how to use such developments to strengthen their position in national-level discussions may gain added clout.

1.3 THEORY BLENDING AND CASE STUDIES

Theory Blending

With this book, I hope to promote and expand the dialogue between various strands of political science and sociology theories on the development of national climate policy. This blending of different scholarly traditions is inspired by the approach developed by pioneers of organization theory, among them Johan P. Olsen, Jim March and Philip Selznick, whose work is just as relevant for sociologists as for political scientists.

In the social sciences, groups of like-minded scholars have dug themselves down into little trenches, focusing solely on their own research interests instead of engaging in the larger scientific discussions about pressing societal issues. As Fligstein and McAdam (2012:208–9) note, 'the demands of successful academic career place an emphasis on a given scholar having a unique empirical and conceptual program of research'. Moreover, scholars are generally encouraged to engage in dialogue only with others of the same theory-persuasion. Theda Skocpol (2003:411) is among those who have warned that such attitudes lead to degeneration of the scientific endeavour, and to the emergence of specialized researchers who learn more and more about less and less.

My objective is not merely to develop a new language that differentiates my chosen approach from others. Readers firmly entrenched in neo-institutional organization studies, institutional work, policy entrepreneurship, international environmental regimes, or multi-level governance will probably be disappointed—this book does not speak directly to on-going discussions within their research communities. Rather, the multi-field framework is pitched to produce dialogue with a range of social science communities, challenging practitioners from all these sub-disciplines to become more involved in studying climate policy.

The multi-field framework consists of various concepts, terms and definitions, many drawing on earlier research. In line with the methodology ideal of comparative politics, it offers a clear ordering of the different components of the framework—which is crucial to analytical productivity (see Collier and Gerring 2009:5; Hall 2003:378).

Case-Study Techniques

This book is more than an exercise in theory: the literature review is combined with comprehensive empirical studies. In line with the recommendations of Bourdieu (1992:233) and Sartori [1970] (2009:25), I have found in-depth empirical studies of climate policies central for deciding which classifications may work, as well as serving as an important source of ideas.

The empirical studies apply the comparative case-study method, which allows for theory pluralism and is conducive to the development of new theory. The method entails detailed examination of a few aspects of historical episodes, aimed at developing historical explanations that may be generalizable to other events (George and Bennett 2005:5). Researchers using the comparative case-study method ask questions about specific sets of cases that exhibit sufficient similarity to be meaningfully compared with one another (Mahoney and Rueschemeyer 2003:8).

Case study is the only method that in and of itself can provide a clear means for identifying mechanisms that take complex causal patterns into account (George and Bennett 2005:21). For many years, the comparative method was widely regarded as a substitute for experimentation and statistic assessment, not a procedure with its own intrinsic value (see Lijphart 1971; Smelser 1973:45; King, Keohane and Verba 1994). Greater recognition of the intrinsic value of the comparative case-study method has led to important improvements in the technique (see Skocpol 2003). At the heart of the comparative method is structured, focused comparison (Hall 2003:378). The method is 'structured' in that the researcher poses the same set of general questions for each case under study, thereby ensuring standardized data collection and enabling systematic comparison (Collier 1993:105; George and Bennett 2005:67). And the method is 'focused' in that it deals only with certain aspects of the cases examined (George and Bennett 2005:67).

Case Selection

Deliberate case selection is central in comparative research. The selection technique is exactly the opposite of the dominant quantitative approach, where randomization is the main point (Gerring 2008:645). Earlier comparative methodology discussions centred on how to select cases that would allow for control over the various 'independent variables' (see Collier 1993:106; Lijphart 1971; Mahoney 2003; George and Bennett 2005:165). All efforts to achieve controlled comparison through case selection implicitly assume that the causal relationships are of a necessary or sufficient nature—and practically all efforts to achieve strict control requirements have failed (George and Bennett 2005:152; Gerring 2008:661). The main reason is that such methodological approaches are poorly suited for capturing probabilistic or more complex kinds of causal explanations, like those that dominate the development of national climate policy.

Several conditions make comparison of the four Norwegian cases selected for study here—carbon capture and storage (CCS), renewable electricity, renewable heating, and energy policy for buildings—particularly suited for a theory-oriented project on national climate policymaking. They represent different climate-policy outcomes, but have a similar structure at the most basic level of explanatory factors: they all started from the same parliamentary decision in 2000; they were all affected by similar strings of EU policy (exposed to rather 'soft' EU climate policy as well as rather stringent EU regulations on state aid); and each had a clear rooting in a specific organizational field.

Cases that do not share such features were rejected; for instance, emissions trading policy has been ruled out because of the complex relationships involving many organizational fields. The similarities in basic structure will, I hope, help to promote a better understanding of the interrelationship between European, national political and national industry–government features. Norway is not an EU member, but, as noted, under the European Economic Area (EEA) Agreement it is obliged to adopt most EU climate policies.

Conditions in the national organizational field and the national political field were particularly important for the selection of cases. Norway's *CCS policy* was promoted by a united political field, but was opposed by the strong and united organizational field of petroleum. Conventional theories on government–industry relationships, such as segmentation and pluralism, led me to expect the organizational field to have had a major impact on CCS policy in Norway. Surprisingly, the actual policy proved to contrast with the positions of dominant field-level actors. Until 2005, *renewable electricity* attracted far less national political attention, and there was more conflict between powerful actors at field level. This case was chosen for study because it seemed very likely that entrepreneurship had played a major role, not least because of demands from the electricity utilities for a shift towards a green certificate system. *Renewable heating* is of interest because of the apparent lack of entrepreneurial engagement: many of the same actors were involved as in the case of renewable electricity, but the policy developed in a remarkably different fashion. As to the final case, *energy policy for buildings,* political support was rather weak, and the organizational field appeared very loosely coupled. Here I expected the organizational field to have exceptionally low importance.

Transparent cases are particularly important for research aimed at developing new theory propositions or shedding light on causal mechanisms. Because the cases in this study have unfolded in Norway, which is a small country with many open sources (document archives, etc.) and easily accessible political and industry leaders, they are also fairly transparent.

Access to information about the causal processes is a precondition for good qualitative research. Further, climate issues gained high political salience in Norway at an early point—even becoming the direct cause of a change in government in the spring of 2000. This level of conflict mirrors developments that we may well expect elsewhere, now that climate policy has emerged as a major political issue of our times. Moreover, the Norwegian situation enables exploration of the outer limits of the power of politicians in relation to climate-policy developments. A further advantage has been my own personal knowledge of the culture, central actors and cases; which has made the Norwegian cases particularly attractive as a backdrop for theory development.

Obviously, choosing cases from only one country may introduce biases into the study. In Norway, it is the petroleum industry that is the most closely linked to the government. In other countries, other industries may have similarly favoured political positions—like the car industry in Germany and Sweden. By contrast, the special relationship between Norway and the EU, through the EEA Agreement and not regular EU membership, is probably of minor importance here. In implementing the EU policies that are relevant to national climate policy, Norway finds itself in much the same situation as most EU member states.

The comparative study combines various tools used in comparative case research. As process tracing and pattern matching are particularly important, I will present each of these below.

Process Tracing and Pattern Matching

Process tracing involves systematic within-case analysis (Hall 2003:397; Mahoney 2003:363; George and Bennett 2005:147). By enabling the researcher to examine whether all the stages in a causal chain fit a certain theory, process tracing serves to enhance reliability and validity. In-depth studies of sequential processes within each case help to specify and identify possible mechanisms at work (George and Bennett 2005:14, 214; Pierson 2004). Process tracing is particularly helpful when explanatory factors and the outcomes are separated by lengthy periods of time (Mahoney 2003:365). This was relevant to all to four Norwegian cases, which were studied over the course of ten years.

As Robert K. Yin (1994) explains, process tracing requires extraordinary rich and diverse empirical foundations. My data compilation has been conducted systematically and similarly in all cases. Relevant publicly available documents, such as parliamentary reports, transcripts of parliamentary debates and governmental decisions, have been important. Various public consultations have been examined: in all, several hundred inputs to four different rounds of public consultations. The inputs have been systemized in tables, which, due to space constraints, are not reproduced in this book.

All cases draw on examination of historical sources that encompass a range of documentation types: history publications telling the story of specific business associations, ministries and agencies; consultancy reports; historical governmental reports; and social scientific research. I have paid particular attention to the structure and culture of key governmental and corporate actors in the organizational fields, examining their organization charts as well as reading their national reports and historical works about their emergence and change over time.

All written correspondence between the Norwegian government and the EFTA Surveillance Agency (ESA) related to the cases explored in the book has been examined. The development in the written dialogue has been mapped and systemized in tables, which, again due to space constraints, are not reproduced in this book. Additional interviews with Norwegian actors as well as EU and ESA officials have proven useful for interpreting the written correspondence.

In all, 95 interviews have been conducted with relevant politicians, civil servants and industry representatives. I have interviewed similar kinds of actors for all cases: high-level politicians from all phases of policy development. For all periods, actors from at least two different political parties have been interviewed, either former ministers or deputy ministers, mainly from the Ministry of Petroleum and Energy, but also the Ministry of Environment, and the Storting (the Norwegian Parliament). I have also interviewed ministerial civil servants, relevant experts in regulatory agencies, representatives from relevant business associations, actors who have followed the issues from the ESA, civil servants in relevant DGs of the European Commission and representatives of Euro-federations.

Because many of the questions probed into highly politically sensitive issues, all interview information has been made anonymous. Many interviewees, and particularly those in political positions, set this as a precondition for being interviewed. To the extent that quotes are given, it should not be possible to identify exactly who provided the information. All interviews have been transcribed, and all interviewees are listed in the interview list. All interviews have been semi-structured: I have had an interview guide and aimed at asking similar questions for all actors, but I also encouraged interviewees to focus on their specific areas of expertise.

Pattern matching comes after process tracing is conducted for each case. George and Bennett (2005:180–204) and Mahoney (2003) advocate variants of pattern matching where certain independent variables and certain outcomes are compared. This involves not only systematic comparison of mechanism and typologies, but also adjustment of the analytical tools along the way. I began by comparing the causal processes involved in the different cases, creating new conceptualizations to capture and explain the development (new typological categories or mechanisms). This entailed comparisons of similar sub-processes in the four cases, active use of existing theory, as well as inventing new terms to capture the causal processes. Next, the new conceptualizations were applied and tested against the same cases, with subsequent revisions as necessary.

Whereas process tracing helps to disentangle the causal processes in one case, pattern matching provides improved possibilities for refining the typologies and mechanisms, requiring the researcher to assess the cases several times. Empirical application of the initial theoretical definitions, concepts and research questions enabled thorough revision and specification of the framework, leading me to adjust the framework itself. Eventually, the theory constructs produced from the case studies were combined into theory framework a more coherent and specified than the initial one. This process was repeated several times.

1.4 PRELIMINARY CONCLUSIONS AND STRUCTURE OF THE BOOK

This chapter has argued that by assessing the policy landscapes (or the multi-field surroundings) in which climate policy is embedded, policy analysts and policymakers can gain a better understanding of climate-policy development. Each climate-policy landscape consists of multiple social systems, or 'fields' as they are called here. This book offers a more coherent understanding of why climate policy emerges and changes, and how various fields influence each other over time—improving their comprehension of the positions and actions of other actors, and giving insights into which policy-shaping factors can be influenced and which that are likely to remain more immune to change. Importantly, the multi-field framework can help climate entrepreneurs—be they politicians, civil servants, industry actors or environmentalists—to overcome some political hurdles related to national policymaking on climate issues. Such knowledge will be crucial to our ability to combat climate change.

The multi-field framework can facilitate comparison of climate policies across countries and sectors, thereby underpinning the further development of accumulative climate-policy research. This book draws on and combines several strands of research otherwise rarely employed in policy studies, with particular weight on neo-institutional sociology. The multi-field framework has been developed specifically for understanding climate policy, but it may prove relevant in assessing other kinds of policy as well.

Part I (chapters 2 to 3) introduces the multi-field framework. Chapter 2 presents theory contributions that focus on societal mechanisms; it conceptualizes the national political field, the national organizational field, and the multi-field European environment, and also specifies a range of societal mechanisms. Chapter 3 discusses entrepreneurship, presenting two main categories of multi-field entrepreneurship: structural and institutional entrepreneurship. This chapter also develops four different entrepreneurial mechanisms.

Part II presents the cases that provide the empirical backdrop. We follow four Norwegian climate-policy developments over the decade from 2000 to 2010. Chapter 4 presents the case of carbon capture and storage (CCS); chapter 5, two cases involving renewable energy: electricity and renewable heating policies; and chapter 6, energy policy for buildings.

Part III presents conclusions, as well as some advice to policymakers. Chapter 7 offers a structured comparison of the four cases, discusses the importance of the various social and entrepreneurial mechanisms and the relative importance of the three fields. Chapter 8 sums up the overarching theory conclusions and finally, chapter 9 presents advice for policymakers and stakeholders.

NOTES

1. Specifically: the 28 current EU member states, as well as the members of the European Economic Area (EEA) (today Iceland, Liechtenstein and Norway). Under the EEA agreement, these countries are obliged to adopt EU market policies, environmental policies and a range of other policies and regulations.
2. Mahoney and Thelen also discuss a fourth category of change, *drift*, referring to changes growing out the failure to adapt and update a policy in order to maintain its traditional impact in a changed environment. That category is not relevant to the discussions in this book, but it may give an apt description of other kinds of climate policy changes than those explored here.

Part I
Theory

Part I

Theory

2   Multi-field Social Mechanisms

2.1 INTRODUCTION

This chapter presents policymakers and policy analysts with tools to map national climate-policy landscapes. It develops a multi-field understanding of policymaking, arguing that we need to take the functioning of complex social systems into account in order to understand how climate policy develops.

The multi-field framework shows how various social systems influence policymaking and policy outcomes. By developing a coherent conceptualization of how multiple fields underpin and influence, constrain and enable national climate policymaking, this book contributes to fill a void in policy studies—a void that needs to be filled if we are to understand the development of climate policy.

The multi-field framework does not seek to determine *a priori* how decision-making unfolds, or what the preferences of decision makers may be. Rather, it facilitates more precise assessments of preference formation and influence distribution in climate-policy developments.

This chapter draws on the work of various branches of sociology, sociological neo-institutionalism in particular, as well as historical institutionalism in comparative politics. Scholars in these traditions have underscored the importance of the meso-level theory of action, specifying the analytical implications of the fact that 'that action takes place between and within organized groups' (Fligstein and McAdam 2012:7). As noted by Fligstein and McAdam (2012:57), 'scholars in a range of disciplines appear to be drawn to the concept and the idea that much of organized life is carried out in constructed social orders'. This chapter discusses the value of the multi-field conceptualization and specifies the distinct societal order of three social systems: the national political field, organizational fields and the European environment. It also conceptualizes the social mechanisms through which multiple fields may influence the development of national policy.

2.2 THE SOCIAL ARCHITECTURE OF FIELDS

What Is a Field?

The founders of the policy studies tradition understood policymaking as constrained and enabled by societal systems (see Wildawsky ([1979] 2007; Lowi 1969). Current policy scholars seldom pay much attention to this. Frank Baumgartner and Bryan Jones ([1993] 2009) are among the relative few who have underscored that social systems—or policy systems, as they call them—underpin and shape policy development. In their studies of US federal policymaking, they specifically state: 'no model of policy change is complete without an appreciation of the multiple venues of policy activity outside of confines of the Washington establishment' ([1993] 2009:xviii).

The multi-level governance perspective increasingly prevalent in EU studies hints at some of the same factors underscored by Baumgartner and Jones (see Hooghe and Marks 2001; Kohler-Koch 1999). Researchers in this tradition use the term 'multi-level' to refer to decision-making in political arenas at different geographical levels that relates to the same issue. They underscore how developments in different social arenas are interlinked and tend to play out in somewhat parallel but uncoordinated ways. Although empirical research on Europeanization has shown that national policymaking in Europe has become increasingly intermeshed with developments within the EU, the multi-level conceptualization has attracted scant attention in national policy studies (see discussions in Kallestrup 2005; Mörth 2003). The multi-field framework takes multi-level governance into account, but adds a systematic exploration of social systems in which the various decision-making 'levels' are embedded.

Johan P. Olsen is probably the political scientist who has most clearly underscored the analytical value of a multi-field approach, although he does not use that phrasing. Olsen argues that within most societal areas we will find a specific configuration of commercial and private organizations that fit into a more or less coherent system of organizations (Olsen 2006:13). Further, he draws the lines back to Max Weber:

> As Max Weber observed, a characteristic of modern society is a differentiation in partly autonomous institutional spheres, such as democratic politics an [sic] governing, the judiciary, market economy, civil society, science, art, religion, family etc.—institutional spheres based upon different, and partly competing, logics of action, structures and processes, normative and causal beliefs, and legitimate resources.
>
> (Olsen 2007 [internet source])

Whereas Olsen prefers the term 'institutional spheres', I have chosen to use 'field'. In their impressive monograph, Neil Fligstein and Doug McAdam (2012:3) present a 'theory of fields', explicating 'an integrated theory that explains how stability and change are achieved by societal actors in circumscribed social arenas'. They show how structural and institutional features of multiple circumscribed social systems –fields—influence and shape human preferences, actions and decisions. Whereas Fligstein and McAdam prefer the term 'strategic action fields', other scholars have applied the terms 'organizational field' or 'institutional field' in similar senses. In this book, I use 'field' as a generic term for circumscribed spheres of political and social life with an identifiable social architecture.

First and foremost, the field perspective on modern societal life is rooted in the neo-institutional sociological literature, where scholars have applied 'fields' as an analytical construct representing the intermediate level between individual organizations and society (Greenwood et al. 2002:58). Dick Scott ([1995] 2008) has argued that this concept is valuable because recognizes and exploits the insight that 'local social orders' constitute the building blocks of today's social systems. It was DiMaggio and Powell ([1983] 1991:64) who provided the first definition of organizational fields: 'those organizations that, in the aggregate, constitute a recognized area of institutional life'. That definition draws heavily on Bourdieu's description of structures and resource distribution within specific spheres which he called 'fields' (see Bourdieu 2005; Bourdieu and Wacquant 1992). Fields are seen as configurations of relations and 'to think in terms of fields is to *think relationally*' (Bourdieu and Wacquant 1992:96, italics in original).

DiMaggio and Powell ([1983] 1991) restricted the term 'fields' to the societal spheres that emerge around specific industries. Their focus was on field-internal homogeneity and unitary makeup, not field internal contestations. Bourdieu, however, regarded fields more as arenas for struggle, and this view has become increasingly influential in the application of the term 'organizational field' (see Bourdieu and Wacquant 1992:101; Wooten and Hoffman 2008). Fields are seen as a dynamic concept, capable of capturing different institutional-structural conditions, varying from homogeneity to heterogeneity, and from consensus to conflict.

Fligstein described fields as 'an arena of social interaction where organized individuals or groups (. . .) routinely interact under a set of shared understandings about the nature of the goal of the field, the rules governing social interaction, who has the power and why and how actors make sense of another's actions' (2008:8). Together with McAdam (Fligstein and McAdam 2012:13), he has later defined strategic action fields as 'meso-level societal orders', as the basic building blocks of modern political/organizational life in the economy, civil society and the state. Fligstein and McAdam argue that fields are socially constructed social spaces involving actors of varying resources, with boundaries that are not fixed but that shift depending on the definition of the issue at stake. They stress that fields are 'constructed' in the sense that they turn on a set of understandings fashioned over time by members of the field (Fligstein and McAdam 2012:10). All field-level organizations will enjoy a certain amount of discretion, and within any field there will be an array of organizations seeking to produce systems of dominance in that space (Fligstein 2001b:15; Leblebici et al. 1991:339).

Most organizations will be involved in several fields—with distinct structures and institutional characters (Hoffmann [1997] 2001:35). Field boundaries differ in permeability: some may be rather open to new members, whereas others will be more closed. In their review of the neo-institutional literature on fields, Wooten and Hoffman (2008) underscore that the structural and institutional patterns of fields will differ, arguing that relations that form around a common technology, say coal production, are not likely to be similar to those relations that form around an issue such environmental protection.

Social systems consist of a range of fields: some have an overarching quality, whereas others are sub-fields within a larger field (Fligstein 2008; Scott [1995] 2008). Moreover, some fields are overlapping, whereas others are exhaustive. The state of a field at any given moment is shaped by dynamics 'internal' to the field *and* by events in a host of 'external' fields with which the field has close ties (see Fligstein and McAdam 2012:58). The myriad of ties to other fields create constraints and opportunities for actors. The social architectures of fields are not fixed: they may shift, adjust and alter over time.

Neo-institutional scholars have been far more interested in how commercial organizations are influenced by informal norms and conventions than how they influence state-led, formal rule-making. Hence, the idea of 'field' has most often been used in reference to the social organization of specific commercial activities, and gradually also in the literature on social movements (see Davis et al. 2005; Fligstein and McAdam 2012). There are some outstanding examples of 'field' literature on environmental issues, such as Andy Hoffmans' (1997 [2001]) book on corporate environmentalism, which offers an institutional account of how the environmental movement has brought about profound changes in the perceptions and practices of large-scale corporations; and Michael Lounsbury's work on university recycling programmes (Lounsbury 2001; Lounsbury et al. 2003). These studies are highly relevant for studies of climate policy, but the explanatory focus is usually on why certain corporate and organizational practices changes, and not on the development of the public policies that govern such changes.

Organization scholars of various kinds have shown greater interest in climate change, but this research is still in its infancy. Bettina Wittneben and colleagues edited a 2012 special issue of *Organization Studies,* exploring organizational change relating to climate change. Summarizing the organizational studies on climate issues thus far, they conclude that, in terms of theory development in organization studies, responses to climate change have been dismal thus far. They note the same tendency as seen in studies of climate policy: that studies often remain 'at the descriptive level with less attention and precision on developing frameworks for understanding the prospects and limits of organizational climate strategies' (Wittneben et al. 2012:1433). Further, they draw attention to the tendency to treat firms as isolated units, divorced from their prevailing social and political context (1435).

Much of the neo-institutional field literature focuses on how fields are created, how they stabilize, and how they may eventually be dismantled (see reviews in Lawrence et al. 2009; Fligstein and McAdam 2012). Policy development is certainly an important element in many of these studies, but neither policymaking nor political steering has been central. Climate-policy development is not primarily about the creation, stabilization and dismantling of fields: it concerns the emergence of and changes in the state regulations that govern the material functioning of fields—so this literature cannot readily be applied in studying climate policy. What is needed is a theory framework for assessing national climate policy that draws on the literature on fields, as well as on the broader political science literature.

A particularly glaring omission in the sociological literature on fields concerns studies of how the political systems pertaining to various sub-issues constrain and enable policymaking, and policy development over time. Political scientists have devoted considerable resources to scrutinizing, describing and comparing national political systems (see Gallagher et al. 2011 for a good overview), but this knowledge has rarely been applied to systematic assessment and comparison of how varying ways of organizing politics affect policy outcomes—not least, the outcomes of climate policy. Moreover, little attention has been paid to the cultural, normative and cognitive dimensions of politics in the literature on comparative politics (see discussions in Culpepper 2011).

Thus, neither neo-institutional sociology nor political science can offer all the tools we need in order to map climate-policy landscapes and understand the forces that shape policy development. As pointed out by Johan P. Olsen (2007 [internet source]), it is unfortunate that 'the two disciplines have, in spite of parallel agendas and many shared assumptions, been in a state of mutual disregard for years'. He calls for conceptualizations that can create bridges between political science and organizational sociology, noting that 'understanding current dynamics in Europe, in particular, requires insight into the tensions and collisions between the major institutional spheres of modern society'. Taking these arguments into account, I draw on insights from both camps when specifying how the national political field, national organizational fields and the European environment may influence national climate policymaking.

The multi-field approach has much in common with the theory of fields presented by Fligstein and McAdam (2012), but differs in three important respects. First, the dependent variable is different. Fligstein and McAdam present a coherent theory of fields so that they can better understand the creation, stabilization and dismantling of fields. In this book, I present a multi-field framework for assessing and explaining how and why national climate policies emerge, evolve and change. Second, the multi-field framework specifies and conceptualizes a field rarely discussed by sociology theorists before: the national political field. In this book, I show that analytical separation between organizational fields and political fields enables us to see more clearly why people in the public administration and elected politicians tend to react differently to similar situations.

Third, I present and specify the analytical term 'the European environment'. Fligstein and McAdam (2012:3) see 'any given field as embedded in a broader environment consisting of countless proximate or distal fields as well as states'. Rather than focusing on one specific field, and treating all other fields as parts of the environment of that field, I distinguish between two core fields and explore how the actors engaged in these fields are influenced by the European environment. True, this is a broad term that includes a great many fields—but it is more constrained and specific than the open, global definition of 'environment' applied by Fligstein and McAdam.

Before specifying the analytical understanding of the two national fields and the European environment, I wish to discuss in greater detail how the two national fields and the European environment can be better understood, by examining their structural and institutional characteristics.

Structures

Pairing classic political science contributions with developments within public administration studies offers a promising avenue for conceptualizing the structural dimension of fields and the European environment. Historically, structural explanations were among the great strengths of political science. However, many political scientists came to overemphasize the extent to which human beings are structurally constrained, exaggerating the control that economic elites exert over the development of formal rules, regulations and compliance mechanisms (see for instance Galtung 1971; Lowi 1969; Lukes 1974). The ensuing criticism led in turn to a focus on lobbying, networks and pluralist competition (Dahl 1961; Rhodes 1997; Sabatier and Jenkins Smith 1993a). However, structural approaches continued to guide assessments of international politics, with many researchers tending to equate national political interests with national industry structures (see Moravcsik 1998). Since the turn of the millennium, more nuanced discussions have emerged about the importance of structural conditions, due not least to the lively debate on Varieties of Capitalism (see e.g. Hall and Soskice 2001; Hancké 2009). While taking these recent developments into account, this book also builds on the valuable insights developed by the first organizational researchers.

Many decades ago, Gulick (1937) described how individual decisions relied on organization structures that shaped the division of work, coordination patterns and hierarchical control mechanisms. Drawing on Gulick, Egeberg (2003:4) has defined an 'organization structure' as composed of formal roles and rules that specify, more or less clearly, who is expected to do what, when and how. Societal fields may be characterized along the same lines, although both inter- and intra-organizational structures are important for the functioning of fields. Various actors, organizations and sub-organizations will seek to shape policy outcomes, but organizations that enjoy favourable structural positions will be more successful than others (Bachrach and Baratz 1962; Lowi 1964; Pfeffer and Salanick 1978). Schatt-schneider has provided a classical formulation of this:

> All forms of political organization have a bias in favor of the exploitation of kinds of conflict and the suppression of others because *organization is the mobilization of bias*. Some issues are organized into politics while others are organized out.
>
> (1960:71, italics in original)

It is not solely the organization of government, but also the organization of private actors, and the structural relationship between the two, that shapes which interests get organized into the policy development processes, and which get 'organized out' (Hall and Soskice 2001). The structures of societal fields create a 'mobilization of bias' that systematically favours some groups over others in the policymaking process (Bachrach and Baratz 1962). Nonetheless, new policy outcomes may create new structures that in turn can undermine, or underpin, prior structural dimensions of fields. No actor will be able to control all the structural resources necessary for achieving a policy outcome, and the distribution of structural resources may vary from one issue to another. Hence, configurations are subject to change over time, although such change will seldom be abrupt or radical (Pierson 2004; Thelen 2003).

Against this backdrop, we can define the *structural* aspect of a field as the laws, contracts, ownership structures, strategic partnerships, organizational charts, public–private partnerships or other written regulations or agreements that create boundaries between different organizations and their interrelationship, and shape how specific organizations, sub-organizations, and actors in specific positions are expected to act. These inter- and intra-organizational features shape the distribution of authority as well as information among actors.

Access to funding is also an important political resource (see Fligstein and McAdam 2012:172). However, and as highlighted by Pepper Culpepper (2011:9), the importance that popular discourse places on money in politics distorts the understanding of the power of commercial actors in most advanced countries. Culpepper argues that companies do have money, and money of course can change minds. However, business often derives most of its political strength from the expertise of managers and their lawyers. In addition, businesses have the authority to take decisions with direct or indirect policy consequences, like investment decisions. Focusing on the distribution of financial resources can make good sense in international politics, or other studies that operate at such highly aggregated levels that the researcher cannot investigate the micro-components of structural power. National policy studies, however, should aim for specific exploration of more tangible micro-factors.

This book will not highlight financial assets as prime structural features. The focus is on the distribution of information and authority as relevant to the development of climate policy. We can start with *authority*.

The *distribution of authority* determines which actors participate and dominate when decisions are made. Although climate policy is a relatively new issue, authority distributions will not be developed *de novo*: historically developed structures will shape the intra-field distribution of authority as well as distribution of authority between fields. Pfeffer and Salanick have highlighted important aspects of authority:

> The final source of control derives from the ability to make rules or otherwise regulate the possession, allocation, and use of resources and to enforce the regulations. In addition to being a source of power, the ability to make regulations and rules can determine the very existence and concentration of power.
>
> (1978:49)

The distribution of authority determines which conflicts will be decided in political arenas and which will be resolved within more private realms (for a general discussion, see chapter 1 in Schattschneider 1960). In most cases, neither the government nor the private realms have full authority: they will generally have authority over different issues. For instance, political arenas tend to make decisions about climate policies, whereas it is industrial actors who decide whether to base their investments on a new policy. How industry actors react to policy decisions will probably feed back into new processes of climate-policy development.

Regulation of authority in an issue-area will 'control the scale of conflict', in the sense that the more actors who are granted authoritative powers, the more conflicts are likely to be activated (Schattschneider 1960:8). When authority is concentrated, one or a few actors will be able to control the decision-making process. Moreover, overlapping or ambiguous authority structures will lead to confusion, tension and conflict (Gulick 1937:9). And here we may note that Europeanization has accorded overlapping competencies to national fields and EU arenas (Hooghe 2001; Hooghe and Marks 2001; Kohler-Koch 1999; Newman 2008).

The policy effects of authority distribution may be counteracted or strengthened by the second structural dimension: the *distribution of information*. Control of information is crucial to the ability of various organizations to influence policy outcomes. As with authority, information is seldom distributed evenly. In most policy issues it is impossible for any single actor or organization to control all relevant information. Nor will organizational actors ever have enough resources to be able to focus on all the information potentially available to them (Simon [1947] 1997). As Schattschneider (1960:136) has argued, nobody knows enough 'to run the government', and those with issue-specific information will be favourably positioned in policy development processes.

Fields have a range of organizations as members, and organizational boundaries create semi-permeable walls that can impede the flow of information (Egeberg 2003:15, Scharpf 1977). Because the flow of information tends to diminish across organizational and sub-organizational boundaries, transmitting information requires substantial coordination (Gulick 1937:33). The larger the organization, the more information it is likely to possess and process: the more labour time an organization controls, the more resources it may designate to acquire, process and present information (Dahl 1961:306). Particularly important here is the number of highly skilled employees and whether they can devote time to policy-directed work. Culpepper (2011:183) sees expertise as a pre-eminent power resource: 'the most relevant currency in convincing state bureaucrats'.

The more complex and technologically demanding an issue, the harder will it be for actors who lack expert information to influence decision-making (Pfeffer and Salanick 1978:48). Developing climate policy requires scientific information of various sorts—from the natural sciences, economics and engineering—so information becomes a valuable source of power. Most decision-making today involves a broad array of various types of expert information. Formal organizational structure will filter what kinds of climate expert information will make it into policy processes. And if there are few alternative sources of information, the information provider will be particularly well-placed to shape policy outcomes.

The distribution of information and authority will determine which field actors engage in and influence the development of climate policy. The distribution of information and authority creates interdependence, with patterns of domination and subordination, as well as various types of coordination between organizations within a field, and between organizations belonging to different fields (Hall and Soskice 2001). Together, the two dimensions influence which fields, and which actors within these fields, will have the upper hand in policy-development processes. Fligstein and McAdam (2012:14) hold that fields can be organized in either a hierarchical or coalitional fashion. With the former, a single dominant group has imposed hierarchical power, whereas the latter opens for looser cooperation between several groups. Once a certain distribution of structural resources has been created, it 'applies to all—those who do not approve as well as those who do' (Pierson 2004:34). On the other hand, how and to what extent various actors and fields actually use their structural positions to influence climate policy will depend on their institutional embedding.

Institutions

An increasing number of policy scholars have recognized that '[h]ow people define their self-interest (their assumptive worlds) depends on culture and history' (Klein and Marmor 2006:908). Culpepper (2011:12–13) holds that political science has paid too little attention to such institutions; further, that political scientists who 'ignore the informal institutions in the economy by assuming they are derivative of formal laws (. . .) will fail even to examine the institutions that matter most'. I will argue that, in order to understand why people support some climate policies and oppose others, we need to understand the informal, cultural features of their fields. Such institutional features influence which decisions policymakers and stakeholders will regard as possible, appropriate and morally acceptable.

Few terms are more commonly bandied about in social science than 'institution'. The word is understood in many different ways, and disagreements and misunderstandings abound as to its analytical importance and qualities. International politics scholars often refer to the features that we here call 'structures' when they use the term 'institutions'. Sociologists generally operate with a broad understanding of institutions, with the features that this book labels as 'structures' as one central component.

Sociologist Dick Scott defines institutions as: '(. . .) comprised of regulative, normative and cultural-cognitive elements that, together with associated activities and resources, provide stability and meaning to social life' (Scott [1995] 2008:48). He specifies that the regulative structures consist of formal rules that are enforced coercively, whereas the normative and cognitive elements are enforced by social, informal reactions. While these specifications are helpful, such a holistic definition also creates some confusion. As Egeberg (2003:10) argues, the 'tendency to classify all kinds of rules, regimes, and organizations as institutional phenomena has given us a poorer concept of institutions'. For better analytical clarity, my analytical framework will restrict the term 'institution' to the normative and cognitive elements described by Scott, whereas the organizational and regulative elements will be regarded as structural features.

The institutional dimension of fields comprises normative elements, like values and norms, as well as cognitive elements, the latter referring to 'the shared conceptions that constitutes the nature of social reality and the frames through which meaning is created' (Scott [1995] 2008:57). Some institutional elements have very general application, such as isolated norms, whereas more elaborate institutional orders create a coherent, interrelated set of identities and rules, such as institutional logics. Both types may play a role in processes of policy development, creating heroes, zeroes and scapegoats by distributing legitimacy and credibility among actors. We now turn to the two institutional features: norms and institutional logics.

Norms create societal obligations and moral expectations of receiving either honour or shame. Here, we will understand 'norms' in line with

Scott's ([1995] 2008:58) description: as related to values and morals. They have a simple composition, highlighting some political ends as more just, more important or otherwise superior to others. Finnemore and Sikkink (1998:895) have shown that, when new norms gain hold, these become the prevailing standard against which new policy proposals are judged. Even though the criteria may not be specified and tangible, or not be applied systematically, actors will use them to validate a specific situation or a specific policy. This will affect which situations become seen as political problems and which ones never enter the political realm (Cohen et al. 1972). After the curbing of carbon emissions emerged as a norm, many actions once regarded as perfectly legitimate have now become morally questionable. That norm clashes with the objective of enhancing economic growth by increasing energy consumption. And climate concerns may clash with the economic profit objectives of some industries (like coal and oil producers), but be more in line with the economic objectives of others (like the producers of renewable energy).

Institutional logics relate more specifically to cognitive elements, such as worldviews, understandings and interpretations. Compliance with cognitive features occurs in many situations because certain ways of doing things are taken for granted, whereas other types of behaviour appear simply inconceivable (Scott [1995] 2008:58). Actors who align themselves with prevailing cognitive beliefs are likely to feel competent and connected, whereas those who are at odds are seen at best as 'clueless' and at worst as 'crazy' (Scott [1995] 2008:59). Different cognitive frameworks will embody different senses of rationality (Thornton et al. 2012: 7). Such cognitive maps may be constraining as well as enabling, and can operate at multiple levels, from the world system to interpersonal interaction (DiMaggio and Powell 1991; Greenwood et al. 2008; Scott [1995] 2008:50).

Norms and cognitive frameworks are inherently linked; together, they contribute to shaping how actors understand the world, their identities, roles and interests (Fligstein and McAdam 2012:46; March and Olsen 1989; Scott [1995] 2008). This is captured in the term 'institutional logics'—a concept applied primarily in sociological studies of changes in specific industry practices, but closely related to the 'logic of appropriateness' as developed by Jim March and Johan P. Olsen (1989; 1998) in political science. According to March and Olsen, actors will act on the basis of how they understand their role, how they interpret the situations they face and which rules they regard as appropriate to follow in such circumstances. Roger Friedland and Robert R. Alford (1991:240) see a systematic relationship between identities and action: institutional rules and understandings will be packaged into specific institutional logics that 'specify the institutional basis for individual and organizational identities, interests and actions'.

Thornton (2004; Thornton and Ocasio 1999; 2008) has specified the concept further, noting that an institutional logic provides a basically coherent template for how to act in different situations. Logics are the bases for action: they shape which issues and problems to attend to, and what answers and solutions are available (Thornton 2004:13–14). Multiple institutional logics are available to individuals and organizations (Friedland and Alford 1991:253); and, while most actors will probably be aware of different logics, they will tend to be more familiar with some (251). Initially, sociologists of the 'neo-institutional' school portrayed institutionalized societal fields as totalizing and shared phenomena—implying that all actors followed the same logic. After applying the institutional logic concept in empirical studies, however, several researchers have identified parallel logics as well as logics in conflict, and it has become more common to argue that social fields may be institutionally inconsistent and contentious (Clemens and Cook 1999; Fligstein 2008; Friedland and Alford 1991; Holm 1995). Moreover, institutional features are never fully determinate: actors will always have some leeway for creativity and experimentation (see Fligstein and McAdam 2012).

The concept of institutional logic has not earlier been applied as an analytical device in policy studies. Later in this chapter, I specify the logics found in organizational fields and in political fields. Organizational fields are populated with actors representing various professions, and the education and training that dominate will vary from field to field. In contrast, political fields are populated by actors seeking to win elections and policy disputes. The logics that drive action in these two fields have different qualities.

Mechanism Variations

I have argued that social orders influence how people act, think and influence the development of climate policy. However, we will need a more precise understanding of the distinct character of the various fields in operation in order to understand how and to what extent different fields influence climate policies. In the following, the systematic differences in the mechanisms through which the various social systems operate and influence climate policy will be discussed. The core argument is that there are systematic differences between the social architecture of the national political field, the organizational field, and the European environment, and so the three social spheres produce different social mechanisms that again will lead to varying climate-policy outcomes.

'Mechanisms' are here understood as the feature governing the policy development process: they are the paths or features that connect causes and effects (Hedström 2008:320, see also Danermark et al. 2002:55; George and Bennett 2005:137 and Hall 2003). Mechanisms are not readily evident through empirical observation: they have to be conceptualized and refined by the researcher (Hall 2003). Each field will influence policy through social mechanisms. Social mechanisms are slow-moving processes, effectuated by actors who follow the institutional patterns in which they are embedded and who act on the basis of their structural positions. Social mechanisms are rooted in specific fields: their functioning will reflect the character of these fields, and they will have structural as well as institutional dimensions. Some mechanisms will be more forceful than others, so the relative importance of a field will depend largely on which mechanisms are in operation.

True, social mechanisms are not the only causal forces involved in processes of policy development. Some actors will be institutionally inventive and act strategically in order to enhance their political clout. I regard such acts as *entrepreneurial*: 'entrepreneurship' refers to acts aimed at enhancing policy influence by altering the distribution of authority and information, and/or activities aimed at altering norms and cognitive frameworks, worldviews or institutional logics. Entrepreneurial mechanisms are discussed in detail in chapter 3.

Several political scientists, among them Theda Skocpol, Paul Pierson, Frank Baumgartner and Bryan Jones, use the term 'feedback' in referring to social mechanisms that unfold over time. In line with these authors, the multi-field framework offers explanations that focus on sequences of events, some of which foreclose certain paths in the development and steer the outcome in other directions (George and Bennett 2005:212; Pierson 2004; Streeck and Thelen 2005). Baumgartner and Jones ([1993] 2009:16–18) distinguish between negative feedback, where an initial disturbance becomes smaller as it works its way through time, and positive feedback, where small disturbances become amplified, causing major disruptions as they operate across time. In this book I develop more detailed conceptualizations of mechanisms, with no *a priori* assumptions about the magnitude of policy change produced by different mechanisms.

In climate policy, we will rarely see only one mechanism operating at a time. The specific context will matter and there will probably be complex interdependencies between various mechanisms; the outcome will depend on the strength of many mechanisms whose strengths again tend to be interdependent (Hall 2003:383; Rueschemeyer 2003:315). Moreover, there may be multiple causal paths and combinations of mechanisms leading to the same outcome (Ragin 1987; Mahoney and Goertz 2006:11). The task of theory development is to discover the different causal patterns that may result in similar policy outcomes, rather than to search for a single, precise and parsimonious explanation (George and Bennett 2005:161).

The only way to advance beyond the boundaries of our knowledge of the causal features that shape climate policy is to make our assumptions and conceptualizations as accurate and explicit as possible, and seek constantly to improve them (George and Bennett 2005:142). That is exactly what I try to do in the following. I specify important mechanisms relating to the three different social systems that may influence national climate policy: the national political sphere, the organizational field and the European environment. For each field, I present four field-specific, policy-shaping mechanisms. Some mechanisms will be more forceful than others: the relative importance of a mechanism will depend partly on the strength of this mechanism and partly on the other mechanism(s) in operation.

2.3 NATIONAL FIELDS AND THE EUROPEAN ENVIRONMENT

Social Mechanisms in National Political Fields

In popular discourse on climate policy, short-sighted and popularity-seeking politicians are often given the blame for the lack of ambitious climate policy. Political science has produced encompassing and impressive literatures on elections and democratization, but has surprisingly little to say about how and when politicians will act as important and independent forces in policymaking (see discussions in Hooghe and Marks 2001, Moravcsik 1998 and Skocpol 1985). Graham Allison (2006:65) holds that policy and public administration scholars have avoided the crucial question of where politics fits into the making of policy. Martin Minogue goes further, arguing that '[t]he greatest problem for most policy analysts is their inability to cope with politics' ([1983] 1993:10). This shortcoming is particularly serious for pressing societal concerns like climate change. Some environmental policy studies have highlighted the importance of leading politicians (see Weale et al. 1996), but we lack a good understanding of the relative importance of political actors. In the following, I propose some possible analytical ways out of the backwater in which we find ourselves, aimed at enhancing our ability to understand the actions of politicians in climate-policy issues.

There is a social science tradition of conflating political and administrative organizations into the term 'state' or 'political-administrative apparatus' (see e.g. Christensen and Lægreid 2001; Fligstein 2008; Fligstein and McAdam 2012; Skocpol 1985). As Pierson (2004:120–21) notes, this presents analytical problems: new governments inherit old administrative apparatuses and have limited ability to alter their organizational structure and institutional-cultural character. Hence, I will analytically separate the policy influence of politicians from that of governmental/state organizations. Government agencies may work as instruments for politicians—but they also have their own dynamics, independent of the political field. In the following, I will specify the social architecture of political fields, and how and to what extent they may influence the development of national climate policy. The role that government agencies play in climate policymaking will be discussed in depth in the section on organizational fields.

Legislative assemblies and committees, political parties, governmental executives and the political leadership of the state ministries concerned with a certain policy area, or sub-issue, take part in a national political field. There will be many arenas for deliberation within and between these units—annual congresses of political parties, parliamentary committee meetings, cabinet meetings and so forth (see Majone 2006:233). Climate policy will be dealt with by varying political fields, such as those pertaining to stationary energy production, building construction, agriculture and transport. These various political fields will produce differing mechanisms that will influence climate policymaking.

Political fields differ from all other societal fields in the sense that the main purpose of such fields is to deal with institutional conflicts in society. The very profession of politicians is to resolve institutional conflicts, giving priority to some values over others, endorsing specific problem diagnoses and developing remedies for appropriate treatment. Whereas other fields relevant to the development of climate policy may focus on producing certain goods or services, the political field produces rules. Because of the special features of the political field, many insights provided by neo-institutional studies that explore organizational field-level phenomena have only limited value. It is primarily political science studies that have underpinned the conceptualization of the mechanisms of the political field, presented in Table 2.1.

Legislature Governing and *Ministerial Governing* are contrasting mechanisms when it comes to policy impact, but before we move on to specify the mechanisms and their influence on policy development, the two dimensions of Table 2.1 will be introduced and discussed more thoroughly.

Earlier in this chapter, the structural dimension was discussed on a more general basis. All fields can be regarded as a nexus of *structural* relationships, but the political field has certain special structural characteristics. First, the distribution of political positions among political parties (the distribution of votes among parliamentary blocs and the composition of the government) delegates authority among political parties. Second, the issue-specific formal distribution of powers between different parts of the government will affect which ministers and which parliamentary committees have a say. Thus, authority is rooted in parliamentary positions as well as in the governmental apparatus—both of which may change with each new election, or as a result of shifting political coalitions. The composition of parliaments and governments changes fairly often, whereas the distribution of authority between certain political arenas tends to be determined by formal rules that are more difficult to change, such as constitutions (Pierson 2004:120–21). Election results may create stable situations with clear majorities, or instable situations with shifting majorities.

The relative importance of the legislative assembly, the government and the ministries will differ from country to country—depending on whether the

Table 2.1 Four political field social mechanisms

Structural pattern ⇨ Institutional pattern ⇩	Concentrated	Distributed
Political Competition	POLITICIZING Intermediate mechanism	LEGISLATURE GOVERNNING Strong policy impact
Garbage Can	MINISTERIAL GOVERNING Weak policy impact	RANDOM DECISION- MAKING Intermediate mechanism

political system is presidential, parliamentarian, federal, or whatever. The distribution of authority may vary from one issue-area to another (Skocpol 1985:17; Egeberg 2003; Baumgartner and Jones [1993] 2009:32). Such differences are evident in the Europeanization literature that has focused on how varying distributions of decision-making abilities—often referred to as 'veto points'—between different parliamentary chambers, ministries or administrative units serve to affect policy outcomes in EU countries (see e.g. Haverland 2000).

In typical European democracies, a cabinet minister is the political head of a major department of state: she or he directs a team of senior civil servants, and has overall responsibility for policy initiation and administration in a key area of state activity (Gallagher et al. 2011:33–37). As only the person in charge of the relevant ministry is in a position to send climate-policy proposals to the cabinet, the minister has a privileged position in the policy jurisdiction in question. However, not all areas of climate policy are governed by one minister only: some may fall under the authority of several ministers, making the distribution of authority less clear-cut. Authority can be said to be more 'concentrated' when the results of an election create a clear majority—consisting of one party or a dominant coalition of parties—with consistent positions.

Turning to the distribution of information in the political field, we note that political parties rarely possess much in-house expertise: they rely heavily on external sources. As Martin Minogue ([1983] 1993:16) explains, for politicians 'information is frequently inadequate or simply not available (. . .) information is a resource, to be used and manipulated'. Because parties in government have access to experts in the administrative apparatus, they tend to have better access to information than parties that are represented only in the legislative assembly. They may also steer which information is available to other political actors. The distribution of information will depend on formal regulations as well as the nature of the issue-area in question. The better the access of politicians in opposition to alternative information sources, the less of a power-tool will this be for the parties in government. Humans can focus on a limited number of issues at a time, so politicians will tend to gather information only about the salient political issues (March and Olsen 1983; Olsen 1983; Baumgartner and Jones ([1993] 2009: xxiii). In consequence, most policy areas are only occasionally the object of political attention.

How will the distribution of structural resources affect how the field influences policymaking? I hold that, if structural resources are concentrated, and are controlled by the political leadership of one state ministry, they are likely to evade political steering altogether. When neither the parliament nor the other political executives are involved in the decision-making, they are also less informed; hence, they are unlikely to evince interest in that political issue. Also the minister responsible will have few incentives for paying much attention, but if she does have an interest in the subject, she will be in a very powerful situation. Things will be radically different if authority and information are spread within the political field: the more political actors who share responsibility for an issue, the more information will the politicians have, with greater incentives for engaging and giving priority to it. The situation is the reverse in organizational fields, where the concentration of structural power enhance the fields' influence over policymaking—a point to which we return later.

Turning to the *institutional* dimension, we may note that politicians tend to give considerable weight to the normative dimension. With respect to climate policy, as with many other issues, politicians will seek to present themselves as holding the morally superior position, whereas their opponents have questionable values. Each political field operates 'according to a different set of rules, with a different agenda, and on different timelines; each responds to different sets of pressures and urgencies; each has its own norms, language, and professional ethos' (Goodin, Rein and Moran 2006:10). The ideology of the political parties as well as historical developments in the policy area in question will influence which issues are regarded as problematic and which solutions seem attractive. Pepper Culpepper (2011:180) holds that many current understandings of politics are premised on the questionable assumption that politics in democracies unfolds through the actions of political parties that assume policy positions with an eye to what they think will appeal to potential voters—but he further specifies this is the case only with issues of high salience: the political dynamics of low-salience issues is very different.

Whether a climate-policy issue is regarded as being of high or low salience will thus have major influence on how it is approached and handled by political elects: if we can understand the significance of an issue, it will be easier to predict how politicians will treat it. I will take the viewpoint of Culpepper further, specifying two different political logics with differing implications for policy development: *competition,* and *garbage can.* Table 2.2 summarizes the implications of the two logics on how political actors will approach and deal with climate policy.

As to the first one, *competition:* many scholars have depicted politics as competitive, often underlining that policy outcomes tend to be aligned to the

Table 2.2 Two political logics

	Political Competition	Garbage Can
Political preferences will reflect	simplified lines of conflict and the action of other political actors	former actions and adjacent developments at specific moments in time
Decision-making is an opportunity for . . .	political victory	symbolic action
Stability of political positions	High	Low
Climate-policy decisions will reflect the positions of . . .	the majority	actors with additional time and energy

aims and strategic calculations of the ruling political parties (see Dahl 1961; Dahl and Lindblom 1953; Olsen 1983; Simon [1947] 1997; Skocpol 1985; Skowronek 1993). However, political competition is not simply a quarrel between different groups that want to make an imprint on policy outcomes. Politicians must take many different concerns into consideration, not the least their chances of getting elected and re-elected (Christensen and Lægreid 2002). Baumgartner and Jones ([1993] 2009:21) argue that issues tend to be drawn into political conflict when a political party sees an advantage in bringing it up for discussion: 'When an issue is drawn into this cleavage, it becomes a partisan issue. The issue is not left to the experts in policy sub-systems; it occupies the major energies of the president and the congressional leaders.' According to Baumgartner and Jones ([1993] 2009:xx), most policy areas will only occasionally be the object of political competition.

As Schattschneider (1960:58) once noted, election dynamics infuse the political field with competition, as *'there is no political substitute for victory in an election'* (italics in original). And Rudolf Klein and Theodore R. Marmor (2006:893) point out that policy decisions will reflect the government's 'larger concerns about gaining (and maintaining) office and doing so legitimately. Uncontentious, even banal as this proposition may be, it is much ignored in the more rationalistic conceptions of policy analysis.'

The main challenge for politicians is to find out what the electorate really wants. This is certainly problematic in respect to environmental issues, where most people do not hold consistent views. People may answer in polls that they want to change their patterns of consumption in order to reduce the environmental impact: but at the same time they may demand higher wages and lower taxes that will allow them to spend more. This lack of consistency makes it hard to know what is really important for the electorate. As Culpepper (2011:7) notes, '[p]oliticians can pay attention to opinion polls to find out public *preferences* on political issues, but they have greater difficulty assessing the political *salience* of an issue: how much the average voter cares about the issue, relative to other issues' (italics in original).

The views of sociologist Harrison C. White (1981) on competitive behaviour in markets can teach us something about the political rationality in issue-areas characterized by political competition. He argued that market actors aim to please the consumer, but because no one really knows how the consumer actually thinks, they must depend on proxies for this. I hold that the politicians do the same thing when aiming for a good election result. Further, White (1981:518) states that market actors define realities and rewards by watching what other 'producers' do. Similarly, politicians will position themselves in relation to other politicians—and not so much in relation to responses from the electorate. Bargaining over political priorities plays out in many arenas within the political field: within political parties, within the state executive, between parliamentary coalitions, and in the media (Allison and Zelikow 1999; Olsen 1983; Pfeffer 1981). Politicians must aim to please several different audiences at the same time (Christensen and Lægreid 2002).

Political competition has three elements. First, politicians will aim to develop clear preferences—and that implies simplifying the lines of conflict. Criss-crossing conflict structures make it intrinsically difficult for politicians to produce clear goal hierarchies (see Christensen 2003; Olsen 1983). All the same, political actors will try to present understandable positions, developed through positioning themselves to the arguments of other political actors. Political discussions will centre on conflicts over objectives and morals, and not complex issues of how to design and implement public policies. In this process, disagreement on how climate policies should be designed will often be portrayed as a yes/no moral issue of whether we should act to combat climate change or not.

Second, the political costs related to unstable positions and changes can be high; instability and shifting stances may signal uncertainty and weak leadership. That makes it attractive for politicians to seek to frame their positions as being stable and consistent over time. Adopting fairly ambiguous positions allows them to alter their decision-making behaviour without appearing to be inconsistent. Because there are limits as to how far or long a politician can stray without being accused of inconsistent behaviour, politicians will refrain from major shifts in their positions. Political positions shaped by competition will tend to be rather sticky, which can help to explain the tenacious character of certain political positions (see Pierson 2004:39–40). Moreover, compromises that enable several parties to present themselves as consistent may be attractive.

Third, political actors like to portray themselves as the winners. Sometimes this can create an incentive for compromises: by taking part in large coalitions, one may create a new majority that leads to shifts in policy outcomes. At other times, competition may take political opponents further away from each other, and the political outcome will shift only in the case of radically different election results.

We may conclude that when the competitive logic dominates a policy issue, the political actors will take into account the arguments and actions of other politicians—all the while trying to present themselves as normatively superior, with stable and reliable positions, and finally as the winners. This behaviour demands considerable energy, so very few issues will be subject to political competition. As Baumgartner and Jones ([1993] 2009:20) note, because legislative politicians will not be able to pay attention to all policy sub-issues all the time, this leads to 'fire alarm oversight': an issue may suddenly become subject to political competition but then quickly fade from the limelight.

Non-salient issues will be governed by a logic of action that unfolds quite differently, in line with what Michael D. Cohen, Jim March and Johan P. Olsen (1972; 1979; see also Kingdon [1984] 2011) have called *garbage can* processes. Such processes have three main characteristics. First, politicians begin with ill-defined preferences. They often will not know what they want until they have seen what they can get; preferences will emerge through action, rather than serving as the basis for action. Political positions will have largely symbolic importance, and neither the politicians themselves nor the electorate will have good long-term memories. Secondly, politicians will not have a deep understanding of the functioning of the policy—or 'technology', as Cohen et al. call it. Politicians will not be aware of the full range of instruments available to them, nor will they be able to understand all the consequences of the various decision alternatives (Goodin, Rein and Moran 2006:10–11).

Third, whether policymakers become involved or not is influenced by many factors: the historical affiliations of their political party, the inputs received when decisions are made, whether the policymakers have a special personal interest in the issue and whether they have time available to devote to a new issue. Timing is essential in this respect: '[w]hat happens is often the almost fortuitous result of the intermeshing of loosely-coupled processes' (Cohen et al. 1979:26). But why did the original authors choose the metaphor of a *garbage can?* They explain that they regarded a policy decision opportunity

> ... as a garbage can into which various kinds of problems and solutions are dumped by participants as they are generated. The mix of garbage in a single can depends on the mix of cans available, on the labels attached to the alternative cans, on what garbage is currently being produced, and on the speed with which garbage is collected and removed from the scene.
>
> (Cohen, March and Olsen 1972:2)

When *political competition* dominates, actors will be acutely aware of the actions of the others within the political field. In contrast, with *garbage can,* politicians will be prone to echo their own prior actions, and will not have the time to follow other actors closely. Further, the competitive logic will lead actors to see decision-making as an opportunity for gaining political victories, whereas decision-making will have only symbolic importance in situations of 'garbage can' logic.

It follows from the above that the stability of political positions will depend on the dominant logic: high in the case of political competition, and low in the case of garbage can. Lastly, climate-policy decisions will reflect the positions of differing actors: the majority will have the main say when political competition dominates, whereas garbage can logic will result in decisions that reflect the positions of those actors who happen to have time and energy available.

Having specified the institutional and structural dimensions, we can now turn to the four different mechanisms that were shown in Table 2.1.

Legislature Governing and *Ministerial Governing* are contrasting mechanisms. Legislature Governing will be at work when formal powers are distributed and politicians compete over the issue in question. Every aspect related to the issue will be subjected to political deliberation, and all actors will do their utmost to win small and large decision-situations relating to the issue. They will follow each other closely; every aspect that can be argued over will be contended. Legislative Governing is a powerful mechanism, in that the political field will be relatively stronger as compared with other fields. Politicians will devote energy to an issue and pay close attention to how it is dealt with in other fields. Because an issue is seen as important, it will be high on the agenda when party programmes are developed, annual meetings are held, government coalitions are formed and state budgets are negotiated. Politicians will play the first violin in climate policy that is driven by this mechanism. However, all environment issues tend to develop in cycles, and no issue will continue to be dominated by this mechanism forever (see Downs 1972).

Radically different is the *Ministerial Governing* mechanism: here, structural power is concentrated, and other parts of the executive government and legislative committees will have hardly any formal authority. The 'garbage can' logic is dominant. Few other politicians are committed to the issue; it is up to the minister with the formal powers to decide whether and how to intervene in a policy area. Due to the lack of political competition on the matter, the minister will normally have little to gain (in terms of popularity) from becoming involved—and will therefore become involved only if she has time to spare, and personal motivation. In practice, this mechanism will tend to give politicians only modest influence on the development of policy. When this mechanism operates in relation to a climate policy, the actual development of the policy in question will probably be shaped by organizational field actors. Due to the lack of political engagement, actual problem-solving is moved downwards, into the ministerial bureaucracy or between civil servants and corporate actors. The minister in charge will tend to be reactive rather than proactive, steering primarily by giving official endorsement to proposals produced by others.

Next, Politicizing and Random Steering are intermediate mechanisms. *Politicizing* will operate in transition periods when new political issues are brought up or old issues are viewed in new ways. It will be at work when structural power in an issue-area is concentrated within one ministry but the issue is gaining in saliency and political competition has set in. Despite their lack of formal powers, wider parts of the political field will engage in the issue, and political parties will compete over it. The opposition will challenge the party/ies in government. Politicians in control of the structural resources will take the decisions—but, as decision-making is supervised by other political actors, they may be influenced by the political competition. The opposition will typically demand greater power over the issue, but whether this will happen will depend on the tactics of the sitting government as well as the opposition—and elections may significantly play into policy development. Some issues can be solved through politicizing, and the issue will shift back into the Ministerial Governing positions—but the conflict can also escalate and shift towards the Legislature Governing situation.

The mechanism of Random Steering will be seen in low-salience issues when many ministries and/or legislative committees share authority over an issue-area, or when formal decision-making authority over an issue is not clearly defined (see Baumgartner and Jones [1993] 2009:32). Political competition will play out in relation to other issues, and few actors will have time and energy for other issues. However, political actors who are not entangled in the salient political discussions may decide to engage in the issue-area, depending on personal motivation. These political actors will probably not be well-coordinated, and the decision outcomes are therefore likely to be incoherent. Note that even though politicians do not control the policy processes when 'garbage can' dominates, the political field may nonetheless be influential (see Kingdon ([1984] 2011). Climate policies under this mechanism will probably be affected by varying political steering signals simultaneously. When we now move on national organizational fields, we will see that their policy impact is determined by other values on the structural and institutional dimensions.

Social Mechanisms in Organizational Fields

Organizational fields may affect climate policy through other mechanisms than political fields. The influence of industry and of societal actors has been discussed at length by political scientists, but conflict runs deep and most scholars have focused more on promoting their own views than developing an analytical language that can foster dialogue between different schools of thought. Drawing on the rich political science and sociology literatures, this section will specify four possible mechanisms for organizational field influence. Climate policies are embedded in fields with very different characteristics—for instance relating to stationary energy production, petroleum exploration, building construction and agriculture. That makes it especially important for climate-policy analysts to develop a clear analytical understanding of how organizational fields influence policy development.

Pluralism and segmentation scholars represent different extremes in the discussion of how national societal groups and governmental organizations and agencies influence policy outcomes. Introducing the term 'organizational field' into policy studies can help us to shed this unfruitful dichotomy, making it possible to capture the various ways that industry–government nexuses can influence the development of climate policy. In the decades after the Second World War, the pluralist view on policymaking gained hold (see Dahl 1961). Pluralists described policymaking as a competition between different societal interests—but they soon encountered considerable resistance from scholars who argued that policy outcomes were controlled by particular segments of societal groups and governmental organizations, and that such relationships were highly resistant to change. Various labels were used to describe such relationships, likes 'corporatism' (Panitch 1980; Schmitter 1974), 'interest-group liberalism' (Lowi 1969), 'segmentation' (Egeberg,

Olsen and Sæthren 1978; Hernes 1983:290; Lieberson 1971) 'iron triangles' (Hernes 1983:291) and 'policy monopolies' (Baumgartner and Jones [1993] 2009:7). All these approaches rest on the assumption that the public is excluded from decision-making, and particularly-favoured interest groups can bypass the normal political forums when policy is made (Lowi 1969:86; Schmitter 1974:100). In this view, it is the experts who steer the development of policy. Some held that governmental organizations were captured or colonized by organized groups, whereas other argued that they influenced each other symbiotically. Despite some internal differences, all these approaches can be grouped under the collective term of *segmentation approaches.*

As a reaction to segmentation, policy theories in the pluralist tradition re-emerged in the late 1970s, and came to dominate policy assessments in the decades to follow. Neo-pluralist approaches see policy outcomes as shaped by free competition and the lobbying activity of many actors (Schmitter 1974:101). The re-introduction of pluralism spawned a whole series of new terms, like 'issue networks' (Heclo 1978), 'policy networks' (Rhodes 1997), 'network governance' (Kohler-Koch 1999) and 'advocacy coalitions' (Sabatier and Jenkins-Smith 1993a). Scholars in this tradition present power as highly decentralized, fluid and situational (Lowi 1964:679). An array of actors is gathered in loose and fairly open networks—journalists, researchers, business representatives, societal actors and administrators (Heclo 1978:88, Sabatier and Jenkins-Smith 1993b:25). Most pluralist works tend to assume that mobilization of one group will lead to the counter-mobilization of another (see Baumgartner and Jones [1993] 2009:4–5).

Neo-pluralists focus mainly on individual preferences and beliefs, not organizational cultures and objectives (see Sabatier and Jenkins-Smith 1993a). Entrepreneurship is deemed highly consequential—particularly when conducted through networking. Some scholars argue that coalitions may be mobilized rapidly and produce surprising new policy outcomes (Gais et al. 1984:163); others expect coalitions to be rather stable in composition, but agree that different coalitions may gain influence at different times (Sabatier and Jenkins-Smith 1993b).

Baumgartner and Jones ([1993] 2009:6) have developed a fruitful middle position. They point out that not all policy areas are segmented all the time: historical factors will 'make them possible at some times and infeasible at others'. Building on this perspective, I argue that two theory constructions represent two empirical extremes: segmentation scholars are interested in situations characterized by the concentration of institutional and structural power, whereas neo-pluralists are focus on situations where decision-making is more open to competing voices and entrepreneurship plays a major role.

The two schools disagree along two dimensions: whether the structural resources are concentrated or distributed among actors, and whether all operate within a certain frame of mind or represent fundamentally different viewpoints. I understand organizational fields as issue- or industry-specific configurations of governmental and private organizations, marked by identifiable structural interrelationships and certain common institutional understandings. The social architecture of fields may vary along the structural as well as the institutional dimensions.

We may expect some processes of the development of national policy to be characterized by segmentation, whereas others will be more pluralist in character. Field boundaries will differ in permeability—but no fields will have entirely fixed boundaries, as projected by segmentation scholars. Neither will the boundaries be wide open, as neo-pluralists assume. I propose seeing segmentation and pluralism as two possible mechanisms thorough which organizational fields influence policymaking. Table 2.3 shows that the two are contrasting mechanisms, but also that two intermediate mechanisms exist.

In terms of *structure,* DiMaggio and Powell ([1983] 1991:65) argued that 'sharply defined inter-organizational structures of domination and patterns of coalition' and a 'certain information load with which the organizations in the field must contend' were key characteristics of organizational field. Later neo-institutionalism scholars showed significantly more interest in the institutional than the structural character of the field, with some exceptions. Together with his colleagues, Scott (Scott et al. 2000:358–60) argues that structural patterns of dominance and subordination—authority—will shape field-level developments. Each field will be governed by specific regularized control: by regimes created by mutual agreement, by legitimate hierarchical authority, or by non-legitimate coercive means. This relates to public laws and regulations, economic agreements between firms, organizational charts, ownership structures and so forth. Whether it is governmental or industry actors that possess the most authority and control of information will vary from field to field. In line with this, Fligstein and McAdam (2012:14) argue that a field may be hierarchically governed, or characterized by collaboration among rather equal organizations; the latter opens for looser coordination and competition between several groups.

In contrast to segmentation scholars and neo-pluralists, Scott, McAdam and Fligstein are sensitive to variation in the structure of fields. Neo-pluralists tend to describe authority structures and information control as intangible, decentralized and shifting (Gais et al. 1984). Conversely, segmentation

Table 2.3   Four organizational field social mechanisms

Structural pattern ⇨ Institutional pattern ⇩	Concentrated	Distributed
One dominant professional logic	**SEGMENTATION** Strong policy impact	**COLLABORATION** Intermediate mechanism
Several professional logics	**TURF BATTLE** Intermediate mechanism	**PLURALISM** Weak policy impact

scholars are inclined to describe structural resources as controlled by a small group of actors, and all inter-organizational relationships are regarded as hierarchically ordered (Schmitter 1974:93). The discussion between pluralists and segmentation scholars is instructive for our understanding of the 'extreme values' on the structural dimension, with heavy concentration of authority and information on the one hand and widely distributed structural resources on the other. In the former situation, all movements will 'go exclusively from the top down'; and in such situations 'the struggle and the dialectic that are constitutive of the field cease' (Bourdieu and Wacquant 1992:102).

Conflicts and discussion will abound if authority and information access are distributed across the field, as in the pluralist view. The organization of the activities of a specific industry will also generate authority-distributing patterns: in industries dominated by a few corporations, authority will be concentrated among a small group of corporate executives, whereas it will be more evenly distributed in industries characterized by many small organizations. Climate-policy positions will tend to be more openly contested in the latter than the former case. Authority may be rooted in a clearly hierarchical structure where one actor has the final say, as with the CEO of a corporation, or a cabinet minister in government. In other cases, authority may be shared among organizations that either coordinate their activities through arguing, bargaining or voting, or simply by continuously making uncoordinated decisions (Fligstein and McAdam 2012; Egeberg 2003:7; Olsen 1983).

The distribution of information may or may not follow the same lines as with authority. To return to climate policy: it is often aimed at re-directing existing practices in industry. Commercial organizations will generally have more information about possible low-carbon solutions and what economic and technological incentives will lead to industry change than will governmental bodies. A government can remain independent of the information provided by industry only if it possesses substantial organizational capacity and in-house technical-economic expertise. In addition, formal requirements may, or may not, give the government access to detailed commercial information. Such levelling-out of the information bias may shift the distribution of power among actors.

What then of the *institutional* character of organizational fields? Early organizational-field studies portrayed the institutional features of these fields as rather unified (see DiMaggio and Powell 1991). DiMaggio and Powell ([1983] 1991:65) argued, somewhat vaguely, that the 'development of mutual awareness among participants in a set of organizations that they are involved in a common enterprise' provided an institutional glue that would eventually lead all the actors in a given organizational field to resemble each other, for instance by adopting similar norms and worldviews. This resembles the arguments of segmentation scholars. For instance, according to Lowi (1964: 689), established expectations and history of earlier decisions will create particular understandings and mind-sets. Eventually, certain professional norms will come to define the 'proper' ways of doing things, and rules of the game that serve established interests will become entrenched (Lowi 1969:92). Both sociological neo-institutionalists and segmentation scholars assumed that the field's origin and the ensuing social processes would shape its institutional hallmarks (Scott et al. 2000; Schneiberg and Clemens 2006; Selznick 1957).

Institutional change, rather than institutional stability, is central to more recent organizational field studies (Lounsbury 2007; Schneiberg and Clemens 2006; Wooten and Hoffman 2008). The introduction of the term 'intuitional logic' (as discussed above) has played a major role in this analytical transition. Several empirical investigations have shown that shifts in institutional logics may radically change the actions of field-level members (see e.g. Greenwood and Suddaby 2006; Schneiberg and Bartley 2001; Thornton 2004; Wooten and Hoffman 2008). This means that if the dominance shifts from one logic to another, field-level actors will change their views on climate policy, perhaps promoting policy measures that they rejected before the shift. However, certain organizational fields, or certain periods in the 'life' of an organizational field, will be characterized by institutional inertia.

The existence of multiple institutional logics enables actors to challenge existing orders as well as propose new policy options (Friedland and Alford 1991:254; Lounsbury 2007:302; Reay and Hinings 2009). For instance, actors who are dissatisfied with a given policy will be strengthened if they can argue that the situation contrasts with the dominant logic, and can propose a new policy measure that fits the dominant logic. Thornton (2004) discusses the institutional logics of many scholarly communities. Building upon her template, this chapter presents and develops three institutional logics relating to climate policy. Different expert communities have tended to favour different approaches. For instance, business economists, economists employed by the government, and engineers generally differ in their views on the challenge of climate mitigation, and therefore in the remedies they propose.

The IPCC contribution on mitigation (Assessment Report 4) offers a foundation for distinguishing three different professional logics that may guide climate policy-induced processes of industrial change (see Metz et al. 2007). These are presented in Table 2.4: (a) market logic, (b) minimizing societal costs, and (c) technical development. Business economists tend to be associated with (a), governmental economists with (b) and engineers with (c). The IPPC authors do not specify the three logics themselves, but the text illustrates the three different views. The three logics will all exist in different variants and will rarely occur in pure form, but it is still possible to isolate their respective coherent, internal structures. Different professional logics will provide different policy recipes and evaluation criteria for measuring policy success.

The market logic (logic a) assumes that commercial organizations possess perfect information and are capable of acting strategically on this information (Fligstein 2001b:13). Firms are expected to strive to maximize their

Table 2.4 Professional logics related to climate-policy development

Type of logic ⇨ Logic components ⇩	Market logic	Minimizing societal cost	Technical development
Objective	Maximizing corporate profits.	Minimizing societal costs.	Enhancing technical development.
Role of commercial organizations	Engaging in strategic competition aimed at maximizing corporate profits	Engaging in all endeavours that yield profit.	Inventing, developing and refining promising technologies.
Appropriate solutions	Market-based support schemes that favour the most profitable low-carbon solutions.	Raising the price of technologies with negative societal effects. Support to projects requiring the lowest level of state aid.	Fostering a broad spectrum of technologies by introducing various technology-specific measures.
Governmental measures	Market measures (e.g. emissions trading and green certificate schemes)	Emissions trading. CO_2 tax. Cost-efficient state funding.	Governmental industry development. Technology standards.

profits in a medium to long-term perspective. In this perspective, governments should work to ensure that low-carbon investments are the most profitable option. Support schemes should be market-based, to encourage market actors to compete in developing the most profitable low-carbon projects (Sims et al. 2007:306). Thereby the 'best' projects will be developed, and actors able to develop the most profitable projects will be rewarded with the greatest profits. Emissions trading systems as well as green certificate schemes for renewable energy are basically designed in line with this logic (see Commission 2005a; 2008a).

Also the logic of minimizing societal cost (logic b) is founded on the assumption of rational economic actors. However, according to this view, corporations will not seek out the most profitable projects, but will engage in all endeavours that can yield profits (Stiglitz and Walsh 2006:158). This logic is aimed at ensuring that conventional industries shift to a low-carbon economy in ways entailing a minimum of societal costs (Gupta et al. 2007:751). Cost minimization is best ensured by emplacing an extra cost on the undesirable effects of conventional production, such as CO_2 emissions. This can be done by introducing a tax or an emissions trading system, although emissions trading designs that allow considerable free allocations will not be in line with this logic. A second-best option is to subsidize investments in low-carbon products. Support schemes should be designed to ensure that such subsidies go only to those projects that require the lowest level of support in order to break even. That in turn necessitates comparing the cost structures of all possible projects, to identify the projects that will need the least support. In this logic, the government has the upper hand, because it determines not only the criteria but also which actors will receive support.

The third logic is based on technological rather than economic criteria. It is assumed that industrial change hinges on technological innovation and its subsequent refinement. Commercial organizations will aim to enhance technological development, and the government should ensure good and stable conditions that enable them to do so. It is the technical quality of the alternatives to conventional production that determines the support levels, so different technologies will receive different levels of support. Further, in this logic, support schemes are designed to ensure long-term stability, so that commercial actors may use the time and resources needed to refine those technologies in which they have greatest expertise (Sims et al. 2007:306). Feed-in tariff schemes that guarantee producers of renewable energy access to the grid, a fixed level of operational support and varying levels of support for different technologies fit well with this approach (Commission 2005a; 2008a). The incentives for competitive behaviour and cost minimization are weak. Technology standards, such as emission limits or energy performance requirements, also fit this logic.

From this discussion and specification of the regulative and institutional features of organizational fields, we can now return to Table 2.3 and discuss the four different policy influence patterns or mechanisms. *Segmentation* will be in operation in fields that resemble descriptions given by segmentation scholars: structural powers are concentrated and largely controlled by one actor, and one professional logic dominates the field. In such situations, 'some policy experts enjoy tremendous freedom of action, seldom being called upon to justify their actions' (Baumgartner and Jones [1993] 2009:8); moreover, 'there tends to be a single understanding of the underlying policy question' (26). The field has a fairly hierarchical structure, with one organization, or a small group of closely aligned actors, at the top. The segmentation and neo-institutional literatures all point in the same direction: that the importance of the organizational field to the policy outcome will be most significant in such situations. There is reason to expect that the field will influence policy the most when this mechanism is in operation. Note, however, that only the views on the dominant actors will leave an imprint on policy development.

In contrast, the *Pluralism* mechanism will be in operation when structural resources are fairly evenly distributed and many different logics are in operation. Various actors will have the opportunity to influence policy development. Due to the many parallel disagreements, actual conflict will not be intense, but minor conflicts are likely to erupt in all kinds of decision-making situations. It is not expected that mobilization of some groups will automatically lead to counter-mobilization, even though that is usual in the pluralist literature. Instead, in situations where few have an intense interest in an issue, the overall organizational field influence on the policymaking may be meagre. To the extent that the field influences climate policy, it will underpin the development of a set of inconsistent policy measures.

Collaboration is an intermediate mechanism; it will operate in fields where structural resources, such as information and authority, are evenly distributed and most participants follow the same professional logic. The actors will tend to develop rather similar viewpoints and be well equipped to understand each other's viewpoints and arguments. There will be good conditions for fruitful collaboration and the level of conflict will tend to be fairly low. This mechanism will create a range of decisions that support the functioning of a certain policy recipe; however, the lack of hierarchically governed coordination may lead to inconsistencies in actor inputs to the development of climate policy.

The *Turf Battle* mechanism is a tougher version of pluralism, where actors disagree strongly, and mobilization is met with counter-mobilization. This mode of organizational field influence will emerge when structural powers are rather concentrated but varying logics are in conflict. The conflict will often be among actors involved in the organizations that have considerable structural resources, as actors with less structural clout will have fewer chances of mobilizing.

Social Mechanisms in European Environment

It is not possible to fully understand political change in European nation-states without taking account of institutional and structural developments within the larger environment (Fligstein 2008; Olsen 2006). Having discussed the functioning of the two national fields in length, we now examine when and to what extent national actors may be overruled by the EU and other European factors. EU policies cover an increasing number of issue-areas, and many studies have highlighted how the authority of the EU is on the rise (see Cowles and Risse 2001:218, Olsen 2006:96; Schimmelfennig and Rittberger 2006). As yet, however, this has led to limited debate on how the EU influences the distribution of power in national policy development (see Börzel and Risse 2006). An important exception from the dominant trend is the volume on Europeanization of national environmental policy from 2005, edited by Andrew Jordan and Duncan Liefferink. Because the EU did not develop an encompassing and significant climate policy until after the Kyoto Protocol was adopted in 1997, we know less about how and to what extent the European environment plays into the development of national climate policies than for other environmental policy areas. Moreover, there have been comprehensive parallel climate-policy developments at the national and the EU levels (see Boasson and Wettestad 2013).

The scholarly literature on European integration and Europeanization can aid our conceptualization of how the EU may affect national policymaking. The literature on 'European integration theory' aims at explaining EU decisions. It has come to comprise a handful of schools, with intergovernmentalism, neo-institutionalism and multi-level governance the most prominent (see e.g. Wiener and Diez [2004] 2009). This literature can provide various insights of relevance to studies of national policy. First, Europe is multi-level, with authority and policymaking influence shared across multiple levels of government—sub-national, national and supranational (Backe and Flinders 2004; Fligstein 2008; Hooghe and Marks 2001; Marks et al. 1996:342). EU as well as other European arenas may intervene in national policy processes. Second, even though national governments are far more powerful than many other European actors, EU policy may under certain conditions constrain or underpin the power of national political executives (see Moravcsik and Schimmelfennig 2009).

The 'Europe hits home' literature—often referred to as the 'Europeanization' literature—assesses national implementation of EU policies (see Mastenbroek 2005; Börzel and Risse 2003). As this research has developed with the objective of understanding differences in formal implementation of EU policies (Börzel and Risse 2006), many contributions assume (implicitly or explicitly) that domestic change will stem from EU policies (see for instance Börzel and Risse 2003; Haverland 2000; Risse et al. 2001:2). However, that is an unrealistic assumption: factors of change may be found internally as well as externally. On the other hand, Europeanization studies have shown that various national factors are involved in both obstructing and promoting national implementation of EU policies. Second, this literature embodies a tension between seeing Europeanization primarily as the transfer of authority from the national to the EU level (e.g. Cowles et al. 2001), and recognizing that the development of EU policy involves complex institutional change processes that unfold at the European level (see Olsen 1992, 2006; Radaelli 2003).

The term 'European environment' as used here comprises all the national and political fields and organizational fields in states that are members of, or otherwise affiliated to, the European Union, EU policy development arenas and European organizations that relate to a specific issue-area. Hence, it takes into account EU-specific factors as well as other developments in EU member states. The European environment is a meta-field, consisting of an array of fields. Despite its vast size, we can still specify its structural and institutional characteristics in different policy areas. The EU has a climate-policy portfolio of varying strength and character in the different areas of climate policy (Boasson and Wettestad 2013). Although it has gained significant structural power in some issue-areas, such as emissions trading, the EU is still without much formal authority in other areas, such as energy policy for buildings. Moreover, climate-policy areas may be dominated by one policy recipe, although most often practices will differ across Europe.

Table 2.5 Four European environment social mechanisms

Structural pattern ⇨ Institutional pattern ⇩	Concentrated	Distributed
One dominant policy model	EU GOVERNING Strong policy impact	BOTTOM–UP HARMONIZATION Intermediate mechanism
Several policy models	UNPREDICTABLE EU GOVERNING Intermediate mechanism	1000 FLOWERS Weak policy impact

Differences in the structural and institutional dimensions will result in different modes of EU influence on processes of national policy development. Table 2.5 presents the mechanisms through which the European environment may affect these processes. What I have termed 'EU Governing' and '1000 Flowers' are contrasting situations, with the EU giving national actors little leeway in the first instance, but letting all—or 'a thousand'—flowers bloom in the second instance. Also the intermediate mechanisms are worth noting, as they influence the development of national climate policy in radically different ways.

The four mechanisms vary along two dimensions: structural pattern and institutional unity. *Structural* elements are central in intergovernmental and multi-level theories on European integration, and have also attracted considerable attention in Europeanization research. Early studies of compliance with EU regulations focused on the coerciveness of these regulations (Mastenbroek 2005). In 2001, Risse et al. provided a definition of Europeanization that underlines the structural power of the EU:

> We define Europeanization as *the emergence and development at the European level of distinct structures of governance,* that is, of political, legal and social institutions associated with political problem solving that formalize interactions among the actors, and of policy networks specializing in the creation of authoritative European rules.
>
> (Risse et al. 2001:3, italics in original)

By highlighting 'formalized interaction' and 'the creation of authoritative European rules', this definition underscores how transfer of authority from the national to the European level gives the EU authority to steer the development of national policy. However, the reference to 'societal institutions' and networks mixes in other factors as well, making that definition less clear-cut.

The emergence of European authority structures often supplements national authority structures, rather than replacing them (Newman 2008:121). This represents a profound shift as regards which actors and concerns get organized *into* national policymaking and which ones get organized *out* (see Schatt-schneider 1960:17–18). Not only will more policy solutions be developed at the European level, but national implementation of EU policy is likely to engage other participants than regular processes of national policy development. Although we might assume that this will reduce the clout of national actors, only empirical investigation can show which national actors lose access to policymaking and which ones may benefit from this change.

European regulatory structures may facilitate cross-field and cross-national coordination of information. Multi-level governance perspectives on EU studies underline how the European Commission, and to some extent the European Parliament, enhance their structural power by creating and governing pan-European expert and policy networks (Eising 2004:218; Hooghe 2001; Hooghe and Marks 2001; Kohler-Koch 1999; Mazey and Richardson 2006). In this sense, the Commission has developed superior access to information. While this may be a more powerful tool for the Commission in the processes of EU-level policy development, it may also contribute to shape the information available to those involved in processes of policy development at the national level.

The structural effects of Europeanization will be stronger when EU organizations are given primary authority to develop national policy requirements, to develop detailed regulations/templates within the policy area and to monitor and facilitate implementation; and when EU regulations are specific and the Commission has coercive follow-up mechanisms available (Boasson and Wettestad 2013; Underdal 2002). As will be discussed later, the EU has required such competencies in only a few areas of climate policy.

Institutional features have attracted increasing attention from Europeanization scholars. Already in 2001, Cowles and Risse (2001:219) pointed out how Europeanization consists of constructing systems of meanings and collective understandings. Radaelli puts institutional processes at centre stage, arguing that 'Europeanization' refers to:

> processes of (a) construction, (b) diffusion and (c) institutionalization of formal and informal rules, procedures, policy paradigms, styles, 'ways of doing things', and shared beliefs and norms which are first defined and consolidated in the making of EU public policy and politics and then incorporated in the logic of domestic discourse, identities, political structures and public policies.
>
> (Radaelli 2003:30)

Radaelli deserves credit for conceptualizing EU policy development as an institutional process, and not as merely a question of static top–down steering. However, he assumes that EU forums and organization will be the prime loci for change, but fails to justify this assumption. It may be argued that EU arenas are only some of many possible arenas in which policy recipes and evaluation criteria may emerge, and that EU policy development will always draw on pre-existing policy recipes provided by member states, European industries or other EU policy areas (see Boasson and Wettestad 2013). Elsewhere Radaelli (2003:30) notes that EU policy will always be 'in the making', and the extent to which EU organizations contribute to construct, diffuse and institutionalize these institutional elements will vary.

The neo-institutional literature indicates that the more a policy recipe diffuses, the more legitimized and theorized will it become (see Finnemore and Sikkink 1998; Meyer 2000; Meyer et al. 1997; Meyer and Rowan ([1983]1991). Policy recipes will start to diffuse because countries mimic each other—the more countries that have adopted a certain model, the stronger will be the pressure for other states to adopt it (Meyer and Rowan ([1983]1991). The cumulative effect of many countries adopting the same policy recipe will create 'peer pressure' among countries (Finnemore and Sikkink 1998:903). Finnemore and Sikkink (1998:900) argue that the emulation (of heroes), praise (for behaviour that conforms to group norms), and ridicule (for deviation) among international actors will contribute the spread of specific norms against which domestic policy is assessed. This discussion indicates that institutional change in the European environment may affect national policy-development processes irrespective of whether the EU or certain states take the lead. The more a specific policy recipe dominates the European environment, the stronger will the pressure be on national policy development.

Let us now move on to the four social mechanisms through which the European environment may influence national climate policymaking, as presented in Table 2.5. As mentioned, the *EU Governing* and the *1000 Flowers* contrast most starkly. The former mechanism will operate when the EU has the upper hand in the policy area, and one policy model dominates. Due to the vast size of the European environment there will always be some variation in how actors regard a policy area, but nonetheless certain policy models may dominate. In such situations, national actors will probably have rather few ideas for alternative climate policymaking, as well as little clout as regards how EU rules and measures are implemented, not least because many implementation decisions are taken in EU organizations and decision-making arenas. For instance, emissions trading has become the dominant model for carbon regulation in Europe, so alternative ways of regulating carbon, like carbon taxation or pollution permit systems, attract little attention. Moreover, the issue is centralized and the EU has the upper hand (see Boasson and Wettestad 2013). The EU can be expected to influence national policymaking the most when this mechanism is in operation—although there will always be room for some national adjustments.

What I have termed the *1000 Flowers* mechanism (based on the dictum *'let a thousand flowers bloom'*) will operate when structural power is spread and several policy models operate in conjunction. This will be the situation for new policy areas not yet covered by EU policy, but it may also occur where the EU has failed to agree on any specific policies and member states have applied panoplies of different policy measures. There will be considerable room for national actors to develop their own policy solutions, without aligning rigidly to EU rules. The EU energy policy for buildings is an example of a climate-policy area where this mechanism is in operation (see Boasson and Wettestad 2013).

Bottom–up harmonization is an intermediate mechanism, occurring when one policy model dominates but the EU has not been granted much formal authority: the harmonization will result from countries that mimic each other, due to peer pressure, competition, or policy trends. There will be considerable national leeway in such situations. The diffusion of renewable energy feed-in support schemes among EU states is a clear example of an area where this mechanism has operated and ensured diffusion of a policy recipe.

Lastly, when the EU has been granted considerable authority but many policy models are in use, we will see *Unpredictable EU Governing*. Due to hefty institutional conflicts, there will be much disagreement over how the EU rules should be interpreted. There will also be considerable room for negotiation between national governments and the EU. Much of the Europeanization literature argues that the increasing volume of EU regulations impoverishes national politicians (see e.g. Cowles et al. 2001). However, that will not be the case when *Unpredictable EU Governing* dominates: potentially the EU has considerable power, but its actual influence on national policy outcomes will rely largely on the strategic capabilities of national actors and their bargaining skills. In climate policy, the EU state-aid regulations for environmental support are perhaps the most lucid example of this situation. The rules for state aid apply *in addition* to the issue-specific policies of the EU: this mechanism may operate in conjunction with the three other European environment mechanisms.

Comparing the Social Systems

Let us now look at the similarities and differences in the national fields and the European environment, and their impact on policy development. Table 2.6 summaries core characteristics of the three social systems.

First, the political field is notable for its lack of ability to deal with many issues simultaneously. A limited number of issues may be on the political agenda at the same time—this is particularly the case for issues that must be dealt with by legislative assemblies. Political debates, hearings and deliberations take time; there is simply not time to deal with many issues in this way. The organizational fields are normally able to deal with a much larger range of decisions; with few people involved in each decision, the decision-making capacity increases. Due to its vast size, the European environment is of course able to handle many issues simultaneously.

Differences in the ability to deal with many different issues at the same time are rooted in structural differences between political and organizational

Table 2.6 Comparing field and European environment characteristics

Kind of field/environment ⇨ Characteristics ⇩	Political field	Organizational field	European environment
Capacity to deal with many issues simultaneously	Low	High	High
Attention to normative aspects	High	Low	Varying
Supply of policy ideas and technical knowledge	Low	High	High
Ability to make pioneering decisions	High	Low	Low
Social mechanism stability	Low	High	High

fields. Organizational field actors generally have varying professional special-izations, and intra- and inter-organizational links ensure a certain degree of coordination between experts. When structural powers are concentrated, one actor will control how all (or nearly all) other field-level actors deal with a given issue. By contrast, many politicians tend to engage in the same political issues: these are those few issues that are deemed salient at a certain time. Few politicians will be motivated to devote much time and effort to issues seen as being of low importance. Hence, low-salience issues will tend to slip away from political field influence; such issues will rather be dealt with by organi-zational field actors. It is easier for a large number of politicians to engage in policy issues underpinned by distribution of structural resources, so political competition will emerge more often in these issues than in issues where the structural power is concentrated. Hence, whereas concentration of structural power will increase the policy influence of organizational fields, power con-centration in the political field will tend to make the field irrelevant.

Concerning institutional features, we can note a clear difference between the two national fields. The political field pays considerable attention to normative issues, whereas the organizational field focuses more on insti-tutional logics. Moreover, the professional logics of organizational fields provide the material from which policies are made, whereas the political logics of the political field will influence how politicians choose the profes-sional logic that is to guide policymaking. Hence, the organizational fields provide the political fields with policy ideas and technical knowledge. Gen-erally, multiple professional and political logics will be at play in the various national political fields and organizational fields involved in the larger Euro-pean environments. However, it is also possible for one professional logic to dominate a policy area throughout Europe, so that other professional and political logics have only marginal importance. In any case, there will always be many sources for new policy ideas within the European environment.

In democratic societies, ultimate authority rests with the political field, so any radical pioneering developments at field level will normally require political endorsement. Fields may change incrementally in a way that eventually leads to path-breaking change, but abrupt change in policy will need political decisions of some kind. Hence the political field has the most ability to make such decisions. Such radical decisions may also be taken at the EU level, but less frequently, due to the complex nature of the EU.

Lastly, the social mechanisms in operation at field level are likely to change far more often and abruptly in the political field than in the organizational field. Regular elections ensure shifts in the structural elements of the political field, and shifting political agendas induce changes in political logic. Organizational fields are more prone to slow-paced shifts, seldom changing much in the course of only one decade, whereas the national political field may change significantly. Also EU steering can change rather swiftly, due to shifts in the Commission, parliamentary elections or changes in the dominant political colouring of the Council.

2.4 CONCLUSIONS

This chapter has presented the complex and many-faceted fields that may underpin and undermine the development of national climate policy. It has provided tools for mapping climate-policy landscapes, each landscape will consist of multiple fields. As explained, a 'field' denotes a circumscribed sphere of political and social life with an identifiable social architecture. Each field has identifiable structural and institutional characteristics and a particular constellation of actors. Drawing on a broad variety of social science literatures, this chapter has specified the distinct societal order of three social systems: national political fields, national organizational fields and European environments.

Each field may operate in four different ways, creating certain field-specific policy-shaping social mechanisms. Social mechanisms are slow-moving processes, effectuated by actors who follow the institutional patterns in which they are embedded and who act on the basis of their structural positions. Social mechanisms are rooted in specific social fields: their functioning will reflect the character of these fields, and they will have structural as well as institutional dimensions. Some mechanisms will be more forceful than others, so the relative importance of a field will depend on which mechanisms are in operation. Some of the 12 mechanisms presented here have been identified by other social scientists previously, but this chapter has gathered and conceptualized them into a coherent analytical multi-field framework.

Policymakers who understand the roles and character of the various fields involved in climate policy will be far better at mapping and understanding the topography of climate policy. Climate-policy landscapes vary significantly across issue-areas, countries and time. However, we have also noted some systematic differences in how political fields, organizational fields and the European environment influence policymaking: the three social systems will influence climate policy through field-specific mechanisms, although the importance of the different mechanisms will vary.

This chapter has focused on each field separately, but in actual policy development they relate to each other throughout the various stages of policy development. Over time, developments in one field may influence which mechanisms are set in operation in other fields. This will be discussed in greater detail in chapter 7, based on systematic comparison of the case studies. We will now move on to chapter 3, which presents how multi-field entrepreneurship may contribute to influence national climate policy.

3 Multi-field Entrepreneurship Mechanisms

3.1 INTRODUCTION

Climate-policy landscapes consist of multiple fields. Mapping the social mechanisms of these fields will enhance our understanding, but it cannot provide the full picture of climate-policy development. It is also necessary to take multi-field entrepreneurship into account. Simply put, some actors are better at influencing political decision-making than others. From the 1980s and onwards, a wave of network approaches has swept policy studies and contributed to vibrant discussions of how particularly skilled actors may influence policy. The role of persuasion in policymaking has gained increased attention as well (Goodin et al. 2006:5). Still, we know little about the relative importance of entrepreneurship in processes of national policy development.

Climate-policy accounts are full of actors with good ideas who manage to achieve extraordinary things. Yet, no matter how politically skilled the actors may be, their climate policy achievements will be constrained by the social conditions of the climate policy landscape. This chapter explores what climate-policy entrepreneurship is and how it may contribute to change policy development. The discussion profits from developments in political science and sociological neo-institutional literatures on entrepreneurship. Although the latter literature does not explicitly discuss entrepreneurship in relation to policy development, it can help us to specify multi-field policy entrepreneurship.

Entrepreneurship is essentially a multi-field activity: international trends create opportunities for entrepreneurship, and the intersection between fields is likely to give rise to entrepreneurship (Battilana et al. 2009:76; Rao et al. 2003; Greenwood and Suddaby 2006; Fligstein and McAdam 2012:28). Because climate policy spans multiple fields, this understanding of entrepreneurship is particularly relevant for our chosen policy area. This chapter does not aim to present the full spectrum of entrepreneurial mechanisms, but concentrates on some common techniques that actors can use in trying to gain greater clout in climate policy.

3.2 CONCEPTUALIZING ENTREPRENEURSHIP

Knowledge Status

Social scientists have described entrepreneurs in various ways: as actors 'essential to the process of policy making' (Roberts and King 1991:147) or 'central figures to the drama' (Kingdon [1984] 2011:189), as actors 'who specialize in identifying problems and finding solutions' (Polsby 1984:171), 'people who seek to initiate policy ideas' (Mintrom 1997:739) 'individuals who change the direction and flow of politics' (Schneider and Teske 1992:737) and actors who 'aim to induce authoritative decisions that would not otherwise occur' (Moravcsik 1999:271). Several writers refer to Schumpeter's definition of innovation, and hold that entrepreneurs innovate by re-combining pre-existing factors or establishing new combinations (see Roberts and King 1991:149; Schneider and Teske 1992).

Neo-institutional sociologists do not focus on specific policy outcomes but seek to explain broader institutional change processes such as the diffusion of neo-liberal ideas and practices. The first neo-institutional works 'assumed away' entrepreneurship: they tended to view institutional features as determinants, but lacked 'an explicit and coherent theory of the individual' (Leca et al. 2006:2). DiMaggio (1988:6) was one of the first neo-institutional writers to raise the issue, arguing that new institutions arise when organized actors with sufficient resources see in them an opportunity to realize certain interests that they value highly. Other neo-institutional scholars have depicted entrepreneurs as actors 'who leverage resources to create new institutions or to transform existing ones' (Maguire et al. 2004:657), 'to whom the responsibility for new or changed institutions is attributed' (Hardy and Maguire 2008:198), who 'break away from scripted patterns of behaviour' (Dorado 2005), actors who mobilize and recombine 'materials, symbols and people in novel and even artful ways' (Hardy and Maguire 2008:206) or 'skilled strategic actors who find ways to get disparate groups to cooperate' (Fligstein 2001a:106).

Thus we see that 'the entrepreneur' as an analytical concept remains rather elusive: some scholars attribute to entrepreneurs all the change that they observe, whereas others simply define entrepreneurs by their formal positions and roles in the change process. Some identify entrepreneurs by their success; others, by their intentions. It has often been implicitly assumed that major determinants of entrepreneurial success lie within the entrepreneur: thus, key elements are the energy and skills exercised by the entrepreneur, and not exogenous factors (see e.g. Dahl 1961:6; Mintrom 1997; Sabatier and Jenkins-Smith 1993a; Schneider and Teske 1992:737). Whereas many policy scholars agree that entrepreneurship has only limited importance, few studies of entrepreneurship have questioned the relationship between entrepreneurship on the one hand and social forces on the other (see Mintrom 1997:738; Roberts and King 1991:172).

Bakir's (2009) discussion on the importance of entrepreneurship in relation to national central banking reforms epitomizes this tendency. He begins by arguing that the importance of entrepreneurs 'depends heavily on the economic circumstances of the time, which are conditioned by the institutional framework within which they operate' (Bakir 2009:578). Later, he goes on to discuss the actual effect of a single entrepreneur, without attaching much weight to structural or institutional factors; and then he ends up attributing to the entrepreneur the most central causal powers. The entrepreneur in Bakir's study is the Minister of the Treasury and Economic Affairs—an individual who controls extensive structural resources and is embedded in an organization with specific institutional characteristics. However, Bakir fails to specify whether the actor is more powerful than indicated by his structural position as such—and so, the extent to which the entrepreneurship of the minister in question actually had an independent and significant effect on the policy outcome remains unclear.

An explicit focus on the paradox of embedded agency underpins neo-institutional discussions on entrepreneurship. The main question has been: if actors are embedded in societal institutions and structures, how can they then be able to envision new practices and subsequently get others to adopt them (see Garud et al. 2007)? Most scholars recognize that 'entrepreneurship involves a degree of dependency on other actors and the resources they control', but all actors are 'subject to regulative, normative and cognitive pressure' (Hardy and Maguire 2008: 199, 207). Some have proposed combining institutional, structural and entrepreneurial forces into one single analytical term: 'institutional entrepreneurship' (see e.g. DiMaggio 1988; Garud et al. 2007). In consequence, some neo-institutional scholars have ended up attributing most structural and institutional changes to the entrepreneur, rather than seeking to grasp the independent effects of entrepreneurship. Thomas Lawrence and colleagues (2009:5) have argued that this literature has tended to 'overemphasize the rational and "heroic" dimension of institutional entrepreneurship while ignoring that all actors, even entrepreneurs, are embedded in an institutionally defined context'.

However, we do find examples of political scientists (such as Dahl 1961; Roberts and King 1991) and neo-institutionalist scholars (like Fligstein 2001a; Lawrence et al. 2009; Leca et al. 2006) who recognize that unsuccessful actors, too, may be entrepreneurs. Moreover, because policy outcomes tend to result from a combination of entrepreneurial efforts and societal mechanisms, a seemingly successful entrepreneur may actually have contributed little to the policy outcome. Persons or organizations that activate the resources provided by their structural position and institutional embedding are merely harnessing social forces—not performing acts of entrepreneurship. For instance, a minister who instructs her civil servants to develop a law in line with her party programme and later endorses the proposal is merely using her structural powers to promote the institutional features in which she is embedded. Distinguishing analytically between the factors that set entrepreneurial acts apart from other social acts, and the factors that make certain kinds of entrepreneurship more successful than others, can help us to assess how and to what extent entrepreneurship influences policymaking.

Thus far, the literature has focused more on success factors than on the defining characteristics of entrepreneurship. The political scientist Robert Dahl (1961) and more recently the sociologist Neil Fligstein (2001a; see also Fligstein and McAdam 2012) have argued that the key to understanding the effect of entrepreneurship lies in concentrating on the individual skills of the entrepreneur. According to Dahl, whether actors are entrepreneurs will depend on how they use their resources and also how '*skillful* or *efficient* they are in employing them' (Dahl 1961:272, italics in original). And: '[s]kill in politics is the ability to gain more influence than others, using the same resources' (p. 307). Similarly, Fligstein (2001a:107) has argued that entrepreneurs are skilled societal actors who will be 'more skilful in getting others to cooperate, manoeuvring around more powerful actors, and generally knowing how to build coalitions in political life'. Further: 'social skill is the ability to take the role of the other in the service of cooperative behaviour' (Fligstein and McAdam 2012:202). Both Fligstein and Dahl indicate that creativity plays an important role, by enabling entrepreneurs to exploit cracks in the societal architecture and find new paths to influence.

However, it is not easy to develop clear definitions of a person's intrinsic entrepreneurial qualities: we cannot measure energy or skills, nor can we assume that actors who perform entrepreneurship in one area always will do so in other social arenas as well. The position of entrepreneur is not a disposition or a quality of an individual: it is a role that becomes available under certain social conditions (Fligstein and McAdam 2012:181). Lawrence et al. (2009:15) argue that if researchers focus on the 'physical or mental effort' involved rather than on tracking down the 'heroic' actors that orchestrated the development, they will be able to advance our understanding of entrepreneurship (or 'institutional work', as they call it). Similarly, the multi-field framework focuses on *entrepreneurship*: it seeks to capture and conceptualize the characteristics of specific acts, rather than the characteristics of specific human beings. This has the great advantage of making it possible to keep track of the actions of persons and organizations. It is much harder—if not impossible—to measure the skills and energy levels of persons and organizations.

Further, studies of industry-level institutional change indicate that entrepreneurship is often a multi-field exercise. Entrepreneurs who are active within several fields will tend to gain new ideas from the different fields in which they operate (see Rao et al. 2003; Greenwood and Suddaby 2006; Fligstein and McAdam 2012:28). Battilana and colleagues (2009:76) argue that the intersection between fields might be more likely to give rise to entrepreneurship. Policy scholars have paid far less attention to the importance of field-spanning actors, but we do find a relevant description in the classic contributions of Schattschneider (1960), describing how the nationalization of US politics created new opportunities for structural entrepreneurship. The transfer of authority from the state to the federal level created the possibility 'for contestants to move freely from one level of government to another in attempt to find the level in which they might try most advantageously to get what they want' (Schattschneider 1960:10–11).

Similarly, Newman (2008) shows that duplication of authority structures in Europe gives societal groups increased possibilities to exert structural entrepreneurship. He notes that '[a]uthority within the European Union is distributed simultaneously across a number of overlapping institutional jurisdictions (. . .) This structure of the European Union opens up access points for policy entrepreneurs' (2008:121). And Tanja Börzel and Thomas Risse (2003:67) maintain that entrepreneurs will use Europeanization as an opportunity to further their goals. Whereas these contributions to the EU literature have focused on societal groups and their increased lobbying on EU policy development processes, in this book I examine how the emergence of EU government structures influences entrepreneurial opportunities in *national* policy processes (see also Backe and Flinders 2004; Hooghe and Marks 2001).

This book takes multi-field entrepreneurship seriously, and specifies entrepreneurial mechanisms that span several fields. That said, a closer reading of the literature on entrepreneurship does make it clear that different authors focus on different entrepreneurship techniques. In the following, I first discuss structural entrepreneurship activities that span several fields, before turning to institutional entrepreneurial efforts.

Structural Entrepreneurship

Structural entrepreneurship involves activities aimed at overcoming the structural barriers to climate-policy influence: obstacles created by the structural distribution of authority and of information. This kind of entrepreneurship will be of relevance to actors who lack formal access to the decision-making arenas responsible for dealing with the climate-policy issues of interest to them, or actors who lack access to information as to when, how and where the issues are to be resolved. For instance, actors outside the government, like environmental groups and business associations, are seldom present when actual climate-policy issues are resolved; they may not have direct access to important decision-makers, like cabinet ministers; they might not even be informed of when the final decisions in legislative committees or the government are to be made.

Entrepreneurship literature often presents networking and agenda-setting as techniques that can be used to overcome such structural challenges. Networking has been highlighted as a means of mobilizing allies and inducing cooperation among others (Hardy and Maguire 2008; Leca et al. 2006). It involves various ways 'to join actors or groups with widely different preferences and help reorder those preferences. (. . .) The trick is to bring enough on board and keep a bandwagon going that will keep others coming' (Fligstein 2001a:114). A 'network' denotes collaboration of a fairly loose, informal and temporary character, and should not be confused with more enduring structural relationships. On the other hand, what starts out as a loose network may eventually become an organization or a formal collaboration between organizations (Leblebici et al. 1991).

Successful networking enables the entrepreneur to coordinate the activity of like-minded actors, get access to important decision-makers and supply decision-makers with new information. Thus it is a tool for diffusing views and motivating others. In order to create large networks, the entrepreneur needs flexibility. Entrepreneurs must be willing to adjust their political projects—and perhaps shift their targets—if that is what is needed in order to get a significant number of actors included in the network (Fligstein 1997; Fligstein and Mara-Drita 1996; Fligstein and McAdam 2012). As Mintrom (1997:739) notes, through networking, an entrepreneur learns the worldviews 'of various members of the policymaking community', and this will enable the entrepreneur to persuade actors with high legitimacy and authority to join the network. Moreover, entrepreneurs can gain access to actual decision-making situations if they include actors with formal authority in their networks (Roberts and King 1991:163). As Hardy and Maguire (2008:2007) explain, '[i]f institutional entrepreneurs do not control rewards and punishment, they can recruit allies that do (. . .) [T]he formal authority of other actors such as the state and professional associations can be harnessed as a resource to support change.' Along the same lines, Mintrom (1997:760) holds that it is particularly important for entrepreneurs to include politicians in their networks.

Considerable attention has been devoted to entrepreneurial *agenda-setting* skills. It is widely acknowledged that entrepreneurship plays important roles in articulating and introducing new policy ideas into the legislative process (see e.g. Finnemore and Sikkink 1998; Kingdon [1984] 2011; Mintrom 1997). Kingdon ([1984] 2011) notes that entrepreneurs will constantly be shopping around in search of decision possibilities where they can succeed in getting their policy ideas on the agenda. Whether and when a political problem becomes coupled with a specific political solution will depend greatly on someone seizing the opportunity to suggest that authoritative, decision-making bodies should link them together (Cohen et al. 1972; Kingdon [1984] 2011). Entrepreneurs will seek to ensure that their 'pet issue' comes up for decision in a situation likely to produce the outcome they favour, and will try to prevent 'their' issue from being included in decision situations seen as ill-suited. If this kind of entrepreneurship succeeds, the venue for decision-making will change or the decision will be made at a point when the entrepreneur can provide the decision-makers with information likely to change their decision-making behaviour.

Institutional Entrepreneurship

Institutional entrepreneurship refers to activities aimed at altering people's norms and cognitive frameworks, worldviews or institutional logics. Such entrepreneurial acts will be of relevance to actors who find that the policy in which they are particularly interested is based on norms or methods that they find to be inappropriate or malfunctioning. The problem is not so much that the decision-makers do not have good information, but that they systematically interpret this information in the 'wrong way', according to the entrepreneurs. Hence, institutional entrepreneurship is about altering decision-makers' preferences and ways of thinking. For instance, persons with one type of professional background—engineers, for example—may disagree with climate-policy measures based on other professional approaches, such as business economics.

Entrepreneur literatures often refer to this entrepreneurial method as *persuasion*. Policy scientists and neo-institutionalists alike see persuasion as central to such entrepreneurial activity. Indeed, according to Robert Goodin and colleagues, 'policy making is mostly a matter of persuasion' (Goodin et al. 2006:5). Furthermore, Finnemore and Sikkink (1998) describe persuasion as the 'mission' of the entrepreneur, and Baumgartner and Jones [1993] 2009:29) hold that 'entrepreneurs use argumentation as a formidable political weapon in their efforts to manipulate political debates'.

Many accounts of institutional entrepreneurship centre on the art of framing (see Goffmann 1974; Snow and Benford 1988). According to Fligstein and McAdam (2012:50–51), the basic challenge for entrepreneurs 'is to frame "stories" that help induce cooperation from people by appealing to their identity, belief, and interests, while at the same time using those same stories to frame action against various opponents'. Newman (2008:120–21) shows that skilful entrepreneurs craft arguments in different ways for different audiences, especially by framing issues so as to overcome possible objections to a proposal. In order to persuade others, the entrepreneur must be able to take into account the perspectives of other actors, and create meanings and frames that appeal to a large number of actors (Fligstein 2001a:106). Entrepreneurs gain power when they are able to 'speak the same language' as the actor(s) they seek to persuade (Bakir 2009). Finnemore and Sikkink (1998:893) argue that, to strengthen their arguments in domestic debates, entrepreneurs will invoke international institutional features, such as norms set out in UN resolutions or EU policies. Friedland and Alford (1991:254) underline that actors who 'are artful in the mobilization of different institutional logics to serve their purpose' will be particularly influential.

We should distinguish between positive and negative framing. *Positive framing* denotes acts directed at presenting certain policy outcomes as good, desirable, legitimate or appropriate. Bernard Leca and colleagues (2006:23; see also Battilana et al. 2009:68) focus on this kind of entrepreneurship when they define institutional entrepreneurship as acts that break with the existing institutions at the field level. Entrepreneurs may try to show their preferred solution is compatible with dominant norms and institutional logics (Greenwood et al. 2002; Leca et al. 2006). Often, the meaning of any particular policy proposal and its link to existing institutional frameworks and logics will not be obvious, but must be actively constructed by the proponents (Finnemore and Sikkink 1998:908).

A specific policy proposal need not appear as appropriate in relation to *all* kinds of institutional logics and norms: it will always be a question of whose assessment is to count in the process of boosting the legitimacy of a proposal (Scott [1995] 2008:60). In order to succeed, the skilled entrepreneur must align to the institutional beliefs of groups that hold prominent structural positions (Finnemore and Sikkink 1998:897, 900). Alternatively, actors may try to change these groups' norms and preferences. Legitimacy may be boosted if it can be proven that the particular proposal is in line with current policy trends. This can be like following the latest fashions: the actors aim at conformity with dominant trends, but they also want to outshine 'the common masses' (Sahlin and Wedlin 2008:223).

Negative framing refers to active de-legitimizing of existing policies and practices. This may entail ridicule and shaming of existing and competing policy proposals—and the actors who defend them (Scott [1995] 2008). James Q. Wilson (1980:370) argued that this may 'put the opponents of the plan publicly on the defensive (by accusing them of deforming babies or killing motorists)'. Finnemore and Sikkink (1998:867) note that state leaders are vulnerable to such strategies, because they are particularly keen to avoid public disapproval and shaming.

Entrepreneurs will seldom develop policy ideas *de novo:* they tend to be inspired by impulses from other fields, often creatively editing and altering these original impulses (see Czarniawska and Sevón 1996). Through such translation, the entrepreneur acts to 'manipulate and reinterpret symbols and practices' (Friedland and Alford 1991:254). The original idea may be changed because only parts of it fit the entrepreneur's own preferences; or the entrepreneur may wish to make the idea appear more attractive to potential allies and coalition partners (Sahlin and Wedlin 2008). Entrepreneurs may also engage in theorizing their preferred policy outcome—specifying, clarifying and generalizing it. The more theorized an idea is, the more general and easily understandable will the aspects attached to it be. Similarly, actual implementation of a policy idea will make it more specified and 'ready to use' (see Røvik 1998; 2007).

Multi-Field Entrepreneurship

Institutional entrepreneurship is directed at altering institutional features in order to lead to policy change, whereas structural entrepreneurship targets policy-decision situations directly. The distinction between institutional and structural entrepreneurship is analytical: in practice, the two techniques may be exercised in tandem, and one actor may perform both kinds of entrepreneurship. Actors will choose the entrepreneurial modes that they see as suited to the situation at hand. For instance, there will not be much point in providing decision-makers with information about a particular policy solution if they find that solution unrealistic, immoral, unnecessary or otherwise dubious. And the converse: why try to persuade decision-makers to endorse a viewpoint if no suitable policy solutions have been presented for that decision making situation? At other times, however, actors may think that only institutional or only structural entrepreneurship will serve for them to influence a process of climate policy.

On this backdrop, this book defines entrepreneurship as acts aimed at enhancing policy influence by altering distribution of authority and information, and/or altering norms and cognitive frameworks, worldviews or institutional logics. This understanding of entrepreneurship opens up for structural as well as institutional entrepreneurship. This definition underlines the purpose of entrepreneurship—to influence policy–but does not maintain that all actors have a strong commitment to a particular policy objective of preferred policy design. Rather, actors who perform entrepreneurship may be motivated by the urge to prove that they are politically skilled, and change their policy objectives and preferences as the policy development unfolds (see Boasson and Wettestad 2013).

By not insisting that entrepreneurs have fixed objectives, my definition differs from how scholars like John Kingdon ([1984] 2011) have understood entrepreneurship. I also have a less specific understanding of purpose than that reflected in the definition of entrepreneurship (or 'institutional work', as they call it) offered by Lawrence and colleagues. They define institutional work as: 'the purposive action of individuals and organizations aimed at creating, maintaining and disrupting institutions'. My definition is also far more narrow, focusing exclusively on policy development—but it also admits the possibility of entrepreneurial acts performed by individual persons as well as organizations or sub-organizations.

My definition of entrepreneurship does not include all actions that may have policy effects. For instance, people in powerful structural positions can be very influential without acting as entrepreneurs: they may change things simply by using their structural position to advance their preferred policy solutions. That said, people in powerful positions may also perform entrepreneurship—for instance, they may seek to change structural or institutional patterns by engaging in networking, agenda-setting, framing or the like. By contrast, actors less structurally advantaged will often be able to influence the development of climate policy only if they succeed as entrepreneurs. Obviously, such actors may have less time and resources to devote to activities directed at influencing policy. Note that merely holding strong opinions about a certain policy-development process is not entrepreneurship; actors must *act* in order to perform entrepreneurship.

One reason for shifting the focus from entrepreneurs to entrepreneurship is to make it easier to differentiate between entrepreneurship and other acts. However, it will still be challenging to distinguish between regular and entrepreneurial acts. In the following, I specify in detail four entrepreneurial mechanisms that may be at work in the field of national climate policy. Note that these do not cover the whole spectrum of entrepreneurship mechanisms, but are only a sample of the ways in which actors may seek to enhance their clout in climate policy.

3.3 FOUR ENTREPRENEURIAL MECHANISMS

Importer Entrepreneurship

Importer Entrepreneurship involves bringing policy features from the European environment into national fields and performing creative translations and interpretations of the current situation. Importer entrepreneurs will refer to developments within the EU in such a way as to frame national conditions in a negative way, while presenting the European alternative as positive. Empirical examples abound (see e.g. Bakir 2009; Kallestrup 2005; Radaelli 2003:36). With respect to climate policy, such entrepreneurs will argue that the EU or some of its member states have developed a good climate-policy solution, one likely to function far better than current national measures.

This type of entrepreneurship emerges in classical *misfit* situations, as described by 'goodness-of-fit' scholars. The 'institutional fit' argument has a strong standing in Europeanization studies. The rationale is that '[t]he degree of adaptational pressure generated by Europeanization depends on the 'fit' or 'misfit' between European institutions and the domestic structures' (Risse et al. 2001:7). Thus:

> Europeanization must be 'inconvenient', that is, there must be some degree of 'misfit' or incompatibility between European-level processes, policies and institutions, on the one hand, and domestic-level processes, policies and institutions, on the other. This degree of misfit leads to adaptational pressures, which constitute a necessary but not sufficient condition for expecting domestic change.
>
> (Börzel and Risse 2003:58)

However, various empirical studies have shown that 'goodness-of-fit' explanations have limited explanatory value (see Mastenbroek 2005 for a discussion). Perhaps one reason is that this school does not provide mechanisms to show how institutional developments within the European environment produce national policy effects. In my view, misfit situations will be effective only if Importer entrepreneurs actively bring EU features into the spotlight, arguing that the EU can provide the solution to given national problems. The 'goodness-of-fit' literature highlights EU regulations as being the carrier of the policy impulse—I hold that also actors performing entrepreneurship may serve as the carrier. Moreover, such actors may be influenced by many actors in the European environment, such as industries, interest groups or experts or certain EU member states. Irrespective of source, the entrepreneur will cite international references in order to enhance the legitimacy of certain ideas or proposals.

European policy recipes and other institutional elements tend to be highly flexible. Only in exceptional instances are they ready-made packages. They generally consist of lightly coupled elements with some new values, 'ways of doing things' procedures or beliefs; and they may well undergo substantial change during the process of policy development (Featherstone 2003:16–17; Mörth 2003:160; Scott [1995] 2008:212). The original European impulse may range from loose ideas and approaches to more specific and tangible policy recipes and technologies (Scott [1995] 2008). Legitimization efforts aim to make an idea appear desirable, proper or appropriate within some specific institutional framework (Scott [1995] 2008:59). Because the meaning of any particular policy proposal and its link to existing institutional frameworks will often not be obvious, the proponents will make reference to European features in order to legitimize their views (Finnemore and Sikkink 1989:908). Importer entrepreneurship entails positive framing of the solutions chosen by the EU and/or other EU countries, coupled with negative framing of the national situation.

The term *Importer* indicates actors performing entrepreneurship. Yes, such actors are inspired by exogenous factors, but they can make a considerable contribution to conceptualizing and shaping how this 'European' model is understood and applied nationally. Neo-institutionalist scholars Kerstin Sahlin and Kjell Arne Røvik have explored how international trends affect national organizational change, highlighting various aspects of institutional entrepreneurship and showing how carriers of international or European trends actively translate European impulses of change (Røvik 2007, 2011; Sahlin-Anderson and Engwall 2002; Sahlin and Wedlin 2008). Moreover, the emerging localization literature in international studies, spearheaded by Amitav Acharya, shows that local agents may 'reconstruct foreign norms to ensure the norms fit with the agents' cognitive priors and identities' (Acharya 2004:239).

Actors who participate in both international and national fields will have particularly good opportunities for transferring the international elements that they favour to the national fields, thereby helping to re-conceptualize the original impulse. Entrepreneurs may act to strengthen the effect of EU policies and other European trends and ideas, but they may also distort, re-shape or weaken the effect of European change impulses. Importer entrepreneurship may lead to the removal of portions of the impulse that the actor dislikes, or it may be altered to fit with the actor's preferences and objectives (Sahlin and Wedlin 2008; Røvik 2011). Acharya (2004) focuses on the important and creative role played by national actors, arguing that normtakers may initiate complex processes, building congruence between transnational norms and local beliefs and practices. Importer entrepreneurship may serve to ensure change as well continuity in national policy. The main message from actors performing Importer entrepreneurship is: *Europe has the solution!*

The aim here is to make a certain policy solution acceptable nationally. The power of Importer entrepreneurship lies in creating a shared understanding of what the ideal policy model looks like, and in ensuring that this becomes taken for granted, not questioned at a later stage. The long-term policy effect hinges on whether the entrepreneur can introduce the model into national policy decision-making. The Importer entrepreneur couples European developments and national policymaking, although that link may be broken at a later stage.

If the entrepreneur succeeds and the model is accepted, future developments in the EU may not be so important any longer. Because the EU lacks a common media agenda, and national policy processes relate primarily to the national field, most participants in the development of national policy lack a detailed understanding of actual developments at the EU level. This may give Importer entrepreneurs good possibilities for depicting the situation in the light they want. Only actors familiar with current developments in the EU can perform Importer entrepreneurship. Whether such opportunities are exploited or not will depend largely on the social mechanisms at work in national fields.

The Fashion Queen

'Fashion Queen entrepreneurship' is the reverse of Importer entrepreneurship as described above. Instead of highlighting the national situation as bad or shameful compared with what goes on within the EU, Fashion Queen entrepreneurs portray their own (national) positions or actions as being at the forefront. This approach combines negative framing of the situation in the EU environment with a positive framing of the national situation. Fashion Queen entrepreneurship presumes a certain *fit* between the European and the national situation. Some national features must be similar to EU institutional or industrial developments—but the national situation must be portrayed as 'better'. This entrepreneurial mechanism draws on a reverse relationship between the national and the European situation from that presented in the 'goodness-of-fit' literature. However, these 'fit' situations can be effective only if Fashion Queen entrepreneurs actively bring EU features into the spotlight.

As with the Importer entrepreneurship, this mode revolves around legitimization, but here the legitimacy of a policy solution hinges on fit. Whereas political scientists have paid scant attention to 'fashion processes', students of organizational change have showed that the diffusion of organization recipes (such as modes of formal organizational structure and leadership philosophies) tends to occur in cycles (Czarniawska and Sevón 1996; Røvik 2012). This has been detected in a whole range of private and public organizations. It is rather like following the latest fashion: actors aim to conform to the dominant trends, but they also want to be somewhat better than the 'common masses' (Sahlin and Wedlin 2008:223). Fashion entrepreneurs aim 'to act differently in order to act in the same way as others' (see Sahlin and Wedlin 2008: 223; Røvik 1998, 2007).

Actors aim at conformity with dominant trends and, simultaneously, at differentiation (Sahlin and Wedlin 2008:223). Adopting policies that appear to be at the forefront of the trend can solve this intrinsic contradiction. The national situation is central: if there is no credible reason for boasting about the national situation, this argument will not work.

The main message here is: *We have the solution for Europe!* This entrepreneurial mode relies to some extent on current developments in the EU. The entrepreneur must be able to produce European actors that leap onto the bandwagon—otherwise the argument will be weakened. Similarly, Fashion Queen entrepreneurs may run into trouble if European actors overachieve. It must still—also over time—be possible for the Fashion Queen entrepreneur to appear as being at the forefront of European trends. Hence, the consequences of Fashion Queen entrepreneurship will depend only partly on the intensity and skills of national entrepreneurs.

Fashion Queen entrepreneurship is invoked in order to defend and enhance a particular national policy development, not to induce new developments. If successful, Fashion Queens will ensure that national developments are connected to larger processes of EU institutional change. As long as there is increasing European competition for the leader position in relation to the policy recipe of the idea in question, the pressure for strengthening the policy measure will rise accordingly. And conversely: if European interest in the policy idea lessens, so will the pressure on improving national performance. In case of entrepreneurial success, the future development of national policy becomes dependent on European dynamics. And because that increases the number of players that may affect the development of policy, this means that the entrepreneur will lose the longer-term ability to affect policy developments.

Shrewd Lawyer

Shrewd Lawyer entrepreneurship emerges as a direct reaction to coercive, structural steering signals from the EU. Deliberation and negotiations are bound to occur in issue-areas where the EU and national actors share authority. Because formal regulations always offer some leeway for interpretation, and because EU regulations tend to be particularly ambiguous, entrepreneurs may succeed in redirecting the steering. If the Shrewd Lawyer entrepreneur succeeds in altering the EU steering, the final agreement will be presented as the work of the EU itself, not of that particular entrepreneur. Any EU steering, irrespective of whether it fits the current national situation, gives national actors a chance to enhance their own power by acting as Shrewd Lawyer entrepreneurs—but this is a possibility available primarily to national-level actors responsible for implementing EU policy.

National entrepreneurs able to include prominent European actors—such as Commission DG officials—in their networks may find it easier to circumvent national actors in prominent structural positions (see Cowles and Risse 2001:229; Newman 2008:105). Instructive here is Mintrom's (1997:739–40) discussion of how entrepreneurs who create networks spanning several US states gain political clout: he shows how this has enabled them to learn more about the policy recipe, and to produce more credible technical inputs to policy development processes at the state level.

The first two entrepreneurial mechanisms presented here—Importer and Fashion Queen entrepreneurship—focus on changing the perceptions and interests of decision-makers. Shrewd Lawyer entrepreneurs, however, target decision-making situations directly. The sociology of law provides a theory foundation for this kind of entrepreneurship. The basic assumption here is that the 'meaning of law and compliance evolves through processes of collective construction and institutionalization' (Edelman 2007). Stinchcombe (2001:182) argues that new interpretations of formal steering tend to emerge when one system of formalized action meets another. The confrontation between national regulation and EU steering gives some actors the opportunity to champion the formalization of their own values and professional logics.

In contrast to the first two modes, Shrewd Lawyer entrepreneurs do not actively take up EU signals and perform the linking between EU policy and the development of national policy. Rather, they happen to find themselves in a situation in which entrepreneurship is needed: they may be forced to act as Shrewd Lawyer entrepreneurs. It is often the actor that receives the EU steering signal that performs this kind of entrepreneurship—not necessarily the national actors with the most at stake. Because ministerial civil servants receive the steering signals, they are the ones who tend to exercise such entrepreneurship—not politicians, or civil society actors. In order to succeed, the actor must recognize that the situation requires entrepreneurship, must understand the thinking that underpins the steering signals from the EU, and must exhibit considerable persuasive abilities. Successful redirection will serve to strengthen the institutional logics and values of the entrepreneur, not those of the EU. Thus, the main message will be: *the EU made us do it!*

Authority will remain with the EU even if Shrewd Lawyer entrepreneurship is successful—the most they can hope for is to alter how this authority is exercised. This entrepreneurial mechanism combines networking, in the sense that national entrepreneurs create network ties to EU actors during the negotiations, with agenda setting, in that these entrepreneurs may influence how the EU rules come to influence the national agenda. The formal outcome of deliberations between national governments and the EU can be reversed only through a new process of negotiation with the European Commission, or, alternatively, by major changes in formal EU rules or practices in the area. The room for entrepreneurship is slim after the case is closed: it is very hard for national actors with other viewpoints than those of the entrepreneurs to redirect the development of national policy at a later stage. Also the entrepreneurs themselves may change their opinions over time— but once the negotiations are over and the formal decisions made, it will be very difficult to change the situation.

Spider

Spider entrepreneurship is a national entrepreneurial mode, exercised independently of developments at the EU level. Actors who employ this mode aim to change the information basis on which formal policy decisions are made, often through a combination of networking and agenda setting. Spiders will act strategically to ensure that their favoured policy solution is dealt with in the most favourable decision-making situation, and/or they will initiate networks that provide policymakers with an information basis that favours this particular policy solution (see Kingdon [1984] 2011; Sabatier and Jenkins-Smith 1993a). Such entrepreneurial networks involve collaboration of a loose, informal and temporary character, and should not be confused with more enduring structural relationships. Moreover, Spider entrepreneurs will tend to produce considerable amounts of written material (see e.g. Mintrom 1997).

The network aspects of information alteration are widely recognized by sociologists as well as political scientists (see e.g. Finnemore and Sikkink 1998; Hardy and Maguire 2008:209; Kohler-Koch 1999; Leca et al. 2006; Sabatier and Jenkins Smith 1993b). Entrepreneurs will create loose networks as tools for diffusing their views, for motivating others and enhancing coordination among the like-minded. In order to create large networks, the entrepreneur must be willing to adjust the political project in question, so as to get a significant number of actors included in the network (Fligstein 1997).

Spider entrepreneurship is primarily structural, but the entrepreneur must possess a certain institutional sensitivity in order to recognize what kind of information may persuade a specific decision-maker. Entrepreneurs gain power when they can 'speak the same language' as the actors they seek to persuade (Mintrom 1997:739). Often, the meaning of a particular policy proposal and its link to existing institutional frameworks and logics will not be obvious, but must be actively constructed by its proponents (Finnemore and Sikkink 1998:908).

More than with any other kind of entrepreneurship, the initiative to Spider behaviour lies with the entrepreneur. Information alteration tends to be initiated by actors that aim for change, not continuity of the policy solutions already in operation. These entrepreneurs will work to downplay existing information sources and highlight their own factual arguments. Their main claim will always be some variant of: *we provide the most credible information!*

Spider entrepreneurship is aimed at influencing formal policy decisions directly, and it works through formal, regulatory change. Some kinds of formal decisions may create long-term effects, but dominant mechanisms within the field or political field will seldom stop operating as a result of a few individual policy decisions. In some instances, Spider entrepreneurship can be highly consequential, with fairly controllable long-term consequences, but the influence will hinge on the conditions within the national fields. Spider entrepreneurs will often have to commit to long-term engagement in order to create long-lasting shifts in field-specific social mechanisms. That makes it difficult for actors to act as Spiders in relation to several issue-areas simultaneously.

Comparing the Four Mechanisms

Table 3.1 summarizes key characteristics of these four kinds of entrepreneurship. We see that the four entrepreneurial mechanisms differ in many ways. They differ in character (two institutional, two structural), they differ in their links to the EU environment (three relate to the European situation;

Table 3.1 Core characteristics of the four entrepreneurial mechanisms

	Importer	Fashion Queen	Shrewd Lawyer	Spider
Character	Institutional	Institutional	Structural	Structural
National–European relationship	Institutional misfit	Institutional fit	Coercion	No relationship
Initiated by	Entrepreneur	Entrepreneur	EU	Entrepreneur
Main message	*Europe has the solution!*	*We have the solution for Europe!*	*The EU made us do it!*	*We provide the most credible information!*
Ways of operating	Entrenching a policy solution in a national field	European institutional competition	European formalization	National formalization
Dominant entrepreneurs	All kinds of actors	Politicians	Civil servants	Marginalized field level actors
Theory foundation	Goodness of fit	Neo-institutional business organization studies	Sociology of law	Windows of opportunity Policy/advocacy networks

one does not), they emerge in different situations (three are initiated by the entrepreneur; one is initiated by the EU), the entrepreneurs vary in their main messages and in their ways of operating, and are based in differing institutional foundations.

Moreover, several entrepreneurial mechanisms will often be exercised in conjunction, and they may be combined in various ways. Note also that this chapter does not present an exhaustive list of entrepreneurship types, but has focused on cross-field entrepreneurship activities. It is challenging to distinguish the effects of entrepreneurship from social mechanisms, but that should not prevent us from trying. Social and entrepreneurial mechanisms operate in conjunction: social mechanisms create entrepreneurial opportunities and influence the extent to which entrepreneurs have a chance of succeeding or not, while entrepreneurship may alter or adjust the social mechanisms that influence a certain policy area.

3.4 CONCLUSIONS

This chapter has examined the role of entrepreneurship in multi-field national policymaking; climate-policy landscapes are, to a certain extent, malleable to entrepreneurship. The chapter has drawn upon entrepreneurship discussions in various social science literatures, and concluded that entrepreneurship can be divided into two categories: structural and institutional. Structural entrepreneurship aims at overcoming the structural barriers to climate-policy influence, that is, barriers created by the structural distribution of authority and of information. Institutional entrepreneurship aims at altering of people's norms and cognitive frameworks, worldviews or institutional logics.

On this backdrop, we have defined entrepreneurship as acts aimed at enhancing policy influence by altering the distribution of authority and information, and/or altering norms and cognitive frameworks, worldviews or institutional logics. Actors who perform entrepreneurship will work to enhance their influence on policy by overcoming or changing the effect of social mechanisms operating within one or several social fields, but they will not necessarily have fixed political objectives: they may well have varying motives.

Drawing on different strands of social science research, this chapter has specified and conceptualized four distinct entrepreneurial mechanisms— here termed *Importer, Fashion Queen, Shrewd Lawyer* and *Spider*. These four entrepreneurship methods differ in many respects, and they will also influence policy development processes in differing ways. By presenting fairly concrete conceptualization of specific mechanisms, I hope to make it possible for climate-policy researchers to identify more easily when entrepreneurship plays important roles in climate policymaking. Three of the four entrepreneurial mechanisms relate to the European environment in one way or the other, so our discussion here has wider implications for the debate on how and to what extent the EU shapes the development of national policy.

Moreover, entrepreneurship may create shifts in the political or organizational fields, resulting in a change from one dominant social mechanism to another—a complex and important issue discussed at length in Part III. Let us now move on to Part II: applying the multi-field framework in empirical case studies.

Part II
Case Studies

4 The Power of Politics
Carbon Capture and Storage (CCS)

4.1 INTRODUCTION

Fossil-fuelled power production is at the centre of the European climate-policy debate, but nowhere has the construction of new power plants given rise to such fierce conflicts as in Norway. At the outset, less than 1 per cent of onshore stationary power production in Norway came from fossil fuel. If the gas-power plant proposals that mushroomed in the late 1990s were to be realized, the situation would change radically. On this backdrop, gas-power construction became one of the most contentious political issues in the country (Tjernshaugen 2007). Eventually Carbon Capture and Storage (CCS) from gas-power plants emerged as a grand compromise that enabled parties with contrasting views on gas power to join in coalition governments. CCS is a suite of technological processes that involve capturing carbon dioxide from the gases discarded by industry and transporting and injecting it into geological formations (Commission 2008a).

In 2000, the majority in the Storting (the Norwegian parliament) argued that introduction of emission trading would create a price on CO_2, and thus no specific governmental engagement in CCS development was needed (St. meld. nr. 54 2001–2002:60). Ten years later, a radical political shift had occurred. Gas power without CCS had been banned and the government has committed itself to investing some 2.8 to 4.5 billion Euro in full-scale CCS. If the CCS project costs prove to be at the high end of the estimate, state funding will have equalled Norway's annual military expenditures (Ministry of Finance 2009; St. prp. nr. 49 2006–2007). This dramatic change in policy has entailed heavy costs, politically as well as economically, and implementation of the measures has led to a series of severe conflicts between the government and industry.

This chapter asks: *Why did the Norwegian CCS policy emerge, and why did it have a Governmental Industry Development character in 2010?* (See Table 4.1). Answers will be sought within the national political field, the national organizational field and the wider European environment. The impact of social mechanisms as well as entrepreneurship will be discussed. The Norwegian CCS policy is a classic 'glass half-full or half-empty' situation. It may be argued that because the petroleum industry, which is targeted by the CCS policy, is the most powerful industry in Norway, we should expect the industry to have a major say in policy development, with the politicians having scant independent importance. In contrast to this conventional wisdom, we note that the political field has succeeded in adopting a policy that contrasts with the wishes of this dominant industry. This puts dominant political science arguments to the test. Neo-realists, rational-choice scholars, neo-Marxist and segmentation scholars all tend to argue that policy decisions will merely reflect the interests of dominant national industries, and that national political actors seldom have much independent impact on policymaking.

However, this case also shows that stark resistance from the segmented organizational field of petroleum can give politicians a hard time in implementing their policy decisions. Persistent opposition from a headstrong field has led the politicians to utilize almost all possible entrepreneurial opportunities to increase their impact. The CCS case has much to teach us about how organizational fields and political fields influence each other over time, and how this dual dynamic shapes policy development. For instance, due to field resistance, the issue has remained dominated by the Legislator Government mechanism for an extraordinarily long period.

The high level of entrepreneurial activity has also meant that the European environment has played more into this case than the dominant European mechanism—1000 Flowers—would lead us to assume. National policy development has been full of references to EU policy, and creative interpretation of the European environment situation has had tangible effects on policy development.

4.2 THE POLICY OUTCOME: GOVERNMENTAL INDUSTRY DEVELOPMENT

The CCS policy falls into the Governmental Industry Development category of climate policy, as discussed in chapter 1. The state has been the motor of the CCS development endeavour, acting as regulator, planner and executor. The public enterprise Gassnova manages the state's direct interests in the various CCS projects and acts as an advisor to the government, but it is the government that takes the decisions on investments, not Gassnova (MPE 2008c). In addition, fossil-fuelled on-shore combustion plants without full-scale CCS are now banned (NVE 2010a).

The Norwegian state has committed to finance CCS in relation to two onshore gas-power plants: one to upgrade the powering of the Mongstad refinery, and the one already operating at the gas processing plant at Kårstø. At Mongstad, the government is involved in two projects, a pilot project (Technology Centre Mongstad, TCM), which is 80 per cent state-owned, and full-scale CCS from the planned gas-power plant. The state will invest

0.6 billion Euro in TCM (St. prp. nr. 38 2008–2009). TCM only captures the CO_2; it does not store it. This is an advanced pilot project, but it does not actually hinder CO_2 emissions.

The full-scale CCS solution at Mongstad is estimated to cost some 3.3 billion Euro, but the credibility of this figure is disputed (StatoilHydro 2009). The Norwegian state is prepared to cover up to 80 per cent of the investment costs, some 2.6 billion Euro. Statoil is required to cover all unexpected costs incurred after the investment decision is made (MPE 2006b). In addition, the government is obliged to pay planning costs up to 0.20 billion Euro (St. prp. nr. 67 2008–2009). Statoil is required to cover the CO_2-related costs they would have had if it were not for the CCS, which corresponds to the price on emissions trading credits. The CCS investment at Kårstø will amount to somewhere between 0.7 and 1.3 billion Euro (NVE 2006; Riksrevisjonen 2010b:151).

The Mongstad TCM is Norway's only CCS project that remains on track. Realization of full-scale CCS has encountered severe problems, at Kårstø as well as Mongstad. In 2009, construction of the Kårstø CCS facility was postponed while the government assessed integration with capture of the substantial CO_2 emissions from the gas processing plant (see St. prp. nr. 67 2008–2009). Then in May 2010 the government decided to postpone the investment decision at Mongstad from 2012 to 2014 (MPE 2010d). If these full-scale projects do proceed as planned, the state will face investment costs of between 2.8 and 4.5 billion Euro, equivalent to Norway's annual military expenditure (Ministry of Finance 2009; St. prp. nr. 49 2006–2007). As long as no investment decision is made, however, it remains unclear whether the state will actually provide the funding. If the CCS projects are realized, the result could be the capture of 4 million tons of CO_2, some 7 per cent of total Norwegian emissions.

The actual CCS policy outcome aims to ensure application of CCS at specific sites, with the winner technology to be selected by the government, not by private actors. The latter are expected to contribute to refining and improving the technology, but it is the government itself that has the leading role in fostering the new technology. We now turn to the process by which this ambitious policy came about, also highlighting how the government has encountered serious problems in ensuring that full-scale CCS is realized.

4.3 FROM POLITICAL GAS-POWER HURDLE TO POLITICAL CCS CONSENSUS

Phase 1, 1995–2000: CCS Enters the Political Scene

To fully understand this case, we need to take a step back into the mid-1990s, when the Labour Party and the majority in the Storting endorsed gas-power development in Norway (Tjernshaugen 2007:48). Naturkraft, owned by the Norwegian petroleum companies Hydro and Statoil and the power producer

Statkraft, launched two gas-power plants, one of them at the Kårstø gas pro-cessing plant site. Initially, promoters argued that Norway's gas-power plants should be comparable with dominant European practices concerning power production; and because gas is cleaner than coal, conventional gas power was hailed as 'progressive' (Tjernshaugen 2007). The environmental move-ment argued in favour of national carbon-emission reductions. If the planned gas-power plants were to be constructed, Norway's carbon emissions would increase by some 10 per cent (St. meld. nr. 29 1997–1998:17). The construc-tion of gas-power plants was depicted as a moral issue, where Norway should set an example for other countries to follow. And the gas-power struggle became the focal point of the Norwegian climate debate.

Initially, the decision-making authority rested within the Ministry of Petro-leum and Energy (MPE), not with the Storting. As politicizing set in, CCS gained attention from the full government. Construction of the first plant was due to start during the 1997 general elections, based on a construction permit issued by the Energy Directorate.[1] However, an intense environmen-tal campaign and strong protests from the Socialist Left Party, the Centre Party (formerly Agrarian), the Christian Democrats and the Liberal Party led the Labour prime minister to instruct Naturkraft to postpone construc-tion until it had been granted a carbon emissions permit (Stortinget 1997; Tjernshaugen 2007:73). This was a new practice: carbon had not been dealt with under the Pollution Act before. Carbon emissions permits were issued by the Environmental Pollution Authority, an agency under the Ministry of the Environment. Thus two ministries were now given formal authority over gas power construction: Environment, and Petroleum and Energy.

The Labour Party experienced disappointing election results, and a minority government opposed to gas power took office in autumn 1997. At this stage, the environmental foundation Bellona was campaigning for CCS, but few took this proposal seriously (Jakobsen et al. 2005:17). Bellona's charismatic leader, Frederick Hauge, had learned about this idea from ide-alistic researchers from within the petroleum industry, who disagreed with the dominant industry scepticism to CCS (Tjernshaugen 2007:116). Erik Lindeberg from SINTEF Petroleum Research was particularly active, reach-ing out to a whole array of environment organizations to promote the idea. Not only was CCS very costly, it was also regarded as a somewhat eccentric and crazy idea. Thus it came as a great surprise when one of Norway's larg-est industrial actors, Hydro (which later merged with Statoil), launched a gas-powered CCS project in 1998 as an alternative to the conventional gas-power plants (Tjernshaugen 2007:96). The Hydro project aroused antago-nism from within the industry but it made CCS appear as a more credible technological remedy for the gas-power hurdle, and Frederick Hauge's calls for CCS were taken more seriously by politicians. The later failure of the Hydro project fuelled the discontent towards CCS among petroleum indus-try actors even further. Technological staff in research institutes and cor-porate R&D divisions in Hydro and Statoil continued to work on CCS technology development, but scepticism remained strong also within the expert communities (see e.g. NOU 2002).

In the 1970s and 1980s, the 'technology development' logic dominated Norwegian petroleum regulation. Back then the petroleum industry as well as Norway's petroleum regulators focused on developing new technologies for offshore drilling, with long-term perspectives to planning and investments (Davis 2006:4, Hanisch and Nerheim 1992; Nerheim 1996). By collaborating with the industry, the government facilitated several major technological breakthroughs. As the global oil market emerged and matured, the petroleum corporations operating on the Norwegian continental shelf found that costly investments reduced their financial value, whereas short-term increased returns were rewarded (Osmundsen et al. 2006:471; Lie 2005). This served to downplay the logic of technology development, and gradually the financial market logic took over (Davis 2006:5). Considerations of financial value came to direct the strategies of the industry, not the long-term economic gains from their operations. Maximizing operational efficiency and short-term earnings was central. Corporations focused on their core competencies, and each business segment was instructed to maximize short-term profits (Antill and Arnott 2004:20; Davis 2006).

Norway's petroleum administration is heavily exposed to the values and belief systems of the industry. Research has shown that the governmental petroleum administration tends to identify with the petroleum corporations (Berge 2005; Boasson 2005). The dominant rationale is that government revenues will continue to grow as long as the petroleum companies increase their profit, and therefore the two have similar interests. The MPE executes most laws that regulate petroleum activity and gas-power production (the Petroleum Law and the Energy Law) and governs the state ownership of Statoil. It is also the line ministry for the regulators of offshore and onshore petroleum activity; it manages the direct ownership of the state in offshore activities (through the fully state-owned limited company Petoro), as well as managing the considerable state ownership in Statoil (MPE 2009b). CCS is a large-scale industrial development idea, at odds with the dominant market institutional logic, and it was strongly opposed by the MPE administration.

The Norwegian Pollution Authority, an agency under the Ministry of the Environment, was dominated by another way of thinking, and the government had not given it any instructions on how to deal with pollution licencing of gas power. In 1999, it granted Naturkraft an emissions permit whereby, in order to be allowed to construct conventional gas-power plants, Naturkraft would have to reduce emissions by 90 per cent, either through CCS or by acquiring emission allowances (SFT 1999). Interview information indicates that this permit was given without direct political steering, but it was in line with the gas-power scepticism in the minority coalition government (Christian Democrats, Centre, Liberals) that held office at the time. As no emissions trading scheme was up and running at the time, this was in practice a CCS requirement. The majority in the Storting, supportive of gas power without CCS, reacted strongly. In March 2000, they agreed:

> Gas power plants in Norway shall be licensed to operate without stricter regulations as to carbon emission than those generally found in other European countries that produce gas power. This majority [in the Parliament] is not familiar with any demands for CO_2 removal in relation to the construction of new power plants within the EEA area. (. . .) This implies that the requirements in the pollution licenses already granted must be altered.
>
> (Innst. S. nr.122 1999–2000, my translation)

At this point, the minority government came under heavy pressure from the majority in the Storting also in many other issue-areas. Because gas power was a good symbolic issue for a resignation debate, the government decided to play it hard. Instead of accepting the instruction from the Storting, the government proposed an alternative formulation: that 'the Pollution Act is not to be weakened as an environmental policy measure' (Tjernshaugen 2007:111). This proposal failed to gain parliamentary support, and so the government resigned. This dramatic backdrop loomed large over the CCS discussions in the years to come, in particular because the outgoing government had interpreted the Storting's decision as 'an attack on Norwegian pollution regulation'. This was a highly creative rhetorical move. In reality, the CO_2 emission permit from the Pollution Authority was far from normal practice, representing instead a significant strengthening of regular pollution control (see Boasson 2005). Normally, the Pollution Authority did not regulate CO_2 at all, and it had no history of requiring large-scale technological change: in most instances it merely required incremental technological improvements.

The strategy of the government placed CCS on a radically different track than other parts of Norwegian climate policy in 2000. Norwegian climate policy at the beginning of the new millennium was dominated by the logic of minimizing societal costs (Boasson 2005; Nilsen 2001; Reitan 1998). According to this institutional logic, only emission cuts entailing the lowest societal costs should be effectuated—whether these were found in Norway or abroad (see Boasson 2005; Tjernshaugen 2007). The government was expected to ensure that low-carbon solutions became more profitable, but not to ensure direct regulation. It followed from this logic that all kinds of double regulation were inefficient (Riksrevisjonen 2010b:26, 169). The argument was that measures introduced in addition to general economic measures would hamper the efficiency of the latter, and therefore the government should refrain from introducing additional measures. Because Norway's petroleum activities were already subject to a CO_2 tax (later included in the EU Emissions Trading System) a CCS requirement in the pollution permit could count as a double regulation. On this backdrop, the petroleum administration and the majority in the Storting argued against pollution regulation of carbon (Boasson 2005; Tjernshaugen 2007). Moreover, this line of argumentation fit well with the dominant market logic that dominated the organizational field of petroleum production in the late 1990s.

The change in government ended the first epoch of CCS in Norwegian history. Even though no substantial CCS policy existed yet, the late 1990s can be regarded as a critical juncture for Norway and CCS policy: decisions made in this period were crucial for the path that the Norwegian CCS policy followed in the 2000s. The organizational field was segmented, and CCS did not fit the dominant field logic, but field-level entrepreneurs were nonetheless able to plant the seeds of a CCS policy in the political field. The politicizing of gas power brought the Storting and the Ministry of the Environment increasing formal powers related to CCS, eventually leading to a Legislature Governing situation in the political field.

Phase 2, 2000–2005: CCS Becomes a Political Objective in its Own Right

European regulations played an interesting role in this period. The European environment was in a '1000 Flowers' situation at the time: few countries had developed any fixed procedures for regulating carbon emissions and few, if any, EU member states regulated carbon under their pollution regulations (Commission 2007b). When the minority Labour government entered office in Norway, it readily transferred the gas-power CO_2 pollution permits from the Pollution Authority to the Ministry of the Environment, shifting authority from the organizational field to the political field. The new government referred to the EU pollution regulation as a reason why carbon should continue to be regulated by the Pollution Act. Through political interference, the politicians now altered the pollution permit, allowing the construction of conventional gas-power plants (Tjernshaugen 2007). Note that the incoming Labour government in March 2000 did *not* change the Pollution Act so that it no longer covered carbon.

More precisely, the government stated that use of the Pollution Control Act in relation to carbon was part of the implementation of the EU directive on pollution prevention and control, the 'IPPC Directive' (Directive 1996). This directive requires *all* emission licences to contain an emissions limit for *all* kinds of pollutants, and that *all* emission limits are to be based on best available techniques. The IPPC Directive defines pollution broadly, as substances from human activity that 'may be harmful to human health or the quality of the environment' (Directive 1996: art. 2). Greenhouse gases are not explicitly mentioned. Thus it was not entirely clear whether carbon emissions should be included in national emission allowances under national pollution regulation acts, and it is evident that the Norwegian Labour government's interpretation of the EU pollution regulation contrasted with the interpretation of other European countries. It also contrasted sharply with the Labour Party's own arguments in the parliamentary debate over gas power half a year earlier. Back then, leading Labour politician Jens Stoltenberg had referred to European pollution regulations in order to legitimize construction of conventional gas-power plants (Innst. S. nr.122 1999–2000). This inconsistency relates to internal Labour Party conflicts over gas power: the youth branch opposed conventional gas power, the labour unions saw gas power as the salvation of Norway's energy-intensive industries, whereas the neo-liberal branch would support gas power if it proved economically profitable. Moreover, the political appointees in the Ministry of the Environment who interpreted EU policy after 2001 were more gas-power sceptical than other factions in the party.

In order to solve the internal conflict, a trade union and Labour Party committee of gas-power opponents and gas-power advocates was set up (LO 2001). This commission ended up by proposing to replace inefficient gas-power plants on offshore installations with power from onshore CCS, and to create state-owned companies responsible for gas grid infrastructure and CCS technology development. One of my interviewees called this outcome 'a solution to a crisis'. The compromise between the two camps met resistance from Prime Minister Jens Stoltenberg, who belonged to the neo-liberal branch of the party, which had not been represented in the committee (Tjernshaugen 2007). Stoltenberg favoured the logic of minimizing societal costs, and regarded the committee's proposal as expensive and far from cost-efficient. In contrast, gas-power proponents among trade unionists anchored their claims in the logic of technology development (see LO 2001). They argued that government should opt for a winner technology, and should assume the economic risks involving in refining the necessary technology. They also held that greater domestic use of Norway's gas resources would sustain activity in national industry and secure jobs.

In practice, the trade unionists re-launched the technology development logic that had dominated Norwegian petroleum regulation in the 1970s and 1980s, before market logic became dominant. After the dominant Norwegian oil corporation Statoil became a publicly traded company at turn of the millennium, the market logic peaked. The state still owned 67 per cent of the shares in the company, but the Norwegian government decided not to interfere with the dispositions of the company (MPE 2009b:21; Statoil 2010). Hence, the organizational field was just as segmented as in the 1995–2000 period.

Initially, gas-power opponents sympathized with neither the industry-development arguments of the trade unions nor their technology development logic (Tjernshaugen 2007). Rather, they argued against gas power on the basis of moral principles, without offering any alternative coherent institutional logic. For many years during the 1990s, Bellona had been the sole grouping to support CCS. Only after CCS emerged as a much-needed political compromise was support forthcoming from among a wider group of environmental organizations who gradually aligned their positions to the logic of technology development. The demarcation lines in the debate shifted. Eventually, erstwhile foes agreed that Norway ought to promote progressive technology development for the best of the whole of Europe—and indeed the world. Moreover, CCS changed from being a compromise, to gaining high normative saliency in its own right.

After the elections in October 2001, the Labour Party government resigned. Immediately prior to that, it had appointed an expert commission to assess CCS policy options further. The new government was a Conservative/Centre minority coalition that included gas-power promoters in the Conservative Party and gas-power opponents in the agrarian Centre Party and the Christian Democrats. The political situation in relation to gas power was murky, with internal disagreement in the political parties, among the parties in government, and between the government and the majority in the Storting. The new government decided to embrace the grand CCS compromise developed by the Labour Party (Semerklæringen 2001; Tjernshaugen 2007:151). Active lobbying from environmental groups facilitated this. At this stage the environmental groups had created strong ties to politicians sceptical to gas power, and interviewees have explained that these groups provided technical information on which the formulations in the governmental declaration were based. Interviewees also stressed that the coalition government needed CCS: it was the compromise that enabled them to form a government.

In 2002 the Commission appointed by the Labour government concluded, somewhat surprisingly, that it was too early to develop full-scale CCS, and merely proposed creation of a public company to facilitate R&D funding (NOU 2002). This was a blow to gas-power promoters and opponents alike. The new government developed a slightly more offensive strategy than recommended by the experts, increasing the funding of CCS research and creating the governmental agency Gassnova (MPE 2005b; St. meld nr. 9 2002–2003). At first, Gassnova was only delegated R&D tasks from the Research Council of Norway, but later it was given a central role in the governmental CCS construction endeavour.

Two years with relative low political focus on CCS followed. Naturkraft had licences to construct two conventional gas-power plants, but because of low electricity prices no decisions on investments were made. Interviews show that under the calm surface, a range of conflicts played out between the government and environmental organizations on the one hand and the petroleum administration and corporate actors on the other. In collaboration with idealistic technical experts from within the petroleum industry, the environmental organizations continually provided politicians with technical information. And in a highly unusual move, the government appointed a senior staff member of Bellona, Thomas Palm, as special advisor to the Ministry of Petroleum and Energy (see also Tjernshaugen 2007). This was a person with CCS competence and strong connections to the engineers who had initially promoted CCS as a climate solution, like the earlier-mentioned Erik Lindeberg of Sintef Petroleum Research.

Apparently, hiring Thomas Palm was not enough to sweep away all the challenges relating to CCS policy development. The Minister of Petroleum and Energy at the time, Einar Steensnæs, described his efforts to work with the administration on CCS like this: 'it is very challenging, and it takes time, and it is evident that internal discussion also taps energy, energy that otherwise could have been used externally' (Tjernshaugen 2007:160, my translation).

The political appointees in the Ministry of the Environment were not confronted with similar opposition. In silence, a crucial decision was made here: gas-power plants should continue to apply for emission permits under the Norwegian Pollution Control Act, even though such plants were now included in the EU Emissions Trading System (ETS) (Ot. prp. nr. 13 2004–2005:45–46). This contrasted with the recommendations from Brussels. When the EU adopted its ETS Directive in 2003, it was decided to exempt carbon from the IPPC pollution regulation regime (Directive 2003). No reference was made to this in the Norwegian decision. Quite the converse: the Norwegian proposal was legitimized with reference to the IPPC Directive's demand for 'coherent regulation of all emissions' (Ot. prp. nr. 13 2004–2005:45).

Interviewees have explained that this creative, entrepreneurial interpretation of EU policy was developed by a group of political advisors to the government. They referred to the important role of the Pollution Act when the government, headed by Prime Minister Bondevik (Christian Democrats), resigned in 2000. This strong symbolic argument silenced the largest party in the coalition, the Conservatives, who at that point still supported conventional gas power. Neither did this move encounter resistance from the Storting. One of my interviewees stressed that, because this was a minority government, the government's decision at this point could have created turmoil in the Storting. But that did not happen because it was 'wrapped up in technicalities, and because it related to EEA juridical issues, nobody would take the chance on any changes'. As we shall see later, this decision was to be a precondition for the ban on the construction of conventional, non-CCS, gas power that was decided at a later stage. Moreover, we see that, at this point, CCS promoters had won the battle over how to interpret and understand the EU pollution regulation in relation to gas power and CCS.

The 2000–2005 phase of CCS policy development was marked by feedback effects from decisions made in the first period, but we can also find some creative interpretations of developments in the European environment, enabling the Norwegian government to regulate CO_2 emissions from gas power plants directly through the Pollution Act. Let us now turn to the political drama that was to unfold after 2005.

Phase 3, 2005–2010: Intense Political Competition Challenging a Segmented Field

The third epoch of the CCS tale began in 2005 when tensions escalated on all fronts during the run-up to general elections in September. First, Naturkraft pronounced its investment decision on the Kårstø gas-power plant. Naturkraft was willing to install CCS, but only on condition that the state covered all the costs (Tjernshaugen 2007:177). Second, Statoil, the biggest industrial actor in Norway, applied for a new large gas-power plant at its Mongstad refinery (see SFT 2006). Since acquiring the petroleum division in Hydro in 2007, Statoil had operated as much as 80 per cent of the oil and gas production on the Norwegian continental shelf (MPE 2010e). Third, the State Petroleum Directorate issued a report showing that CCS for Enhanced Oil Recovery (EOR) was not economically viable. EOR is a technique with the potential for dramatically enhancing the volumes of oil that can be produced from oil reservoirs. The various environmental organizations countered by issuing reports that came to the opposite conclusion. Bellona even argued that CCS development coupled with EOR would, in the long term, provide the Norwegian state with substantial profits (Jakobsen et al. 2005). The latter argument was embraced by the entire political field (Tjernshaugen 2007).

The Labour Party, the Socialist Left and the Centre (formerly Agrarian) Party gained a majority in the parliament in the elections in 2005. Interviews show that CCS was the issue to which these parties devoted the most time and energy in negotiating the political foundations of the new coalition government. According to one interviewee: 'The gas-power issue is a matter of principle that has become sticky in Norwegian politics; none of the parties could afford to lose face in relation to this issue.' Interviewees confirm that information from the environmental foundations Bellona and ZERO had a strong impact on negotiations in 2005. Eventually the new government agreed that, by means of economic measures and promotion of new technology, they would ensure that new gas-power licenses were to be based on CCS; state-owned companies for capture of CO_2 and management of the 'value chain' for CO_2 would be created and state aid made available for installing CCS at Kårstø (Soria Moria 2005:59). This led both the Socialist Left Party and the environmental movement to conclude that the 'fight against polluting gas power is over' (Tjernshaugen 2007:183).

But they were wrong. The gas-power plant planned at Mongstad had not been explicitly discussed in the initial negotiations. Prime Minister Stoltenberg later said that this was not because he had not been aware of this, but rather because he wanted to create a compromise text that would not obstruct the development of gas power at Mongstad (Tjernshaugen 2007:186). Soon a heated discussion erupted concerning CCS at Mongstad. In autumn 2006, the government was seething with internal disagreement over the criteria in the emissions permit. In parallel the government negotiated the emissions criteria with Statoil (Tjernshaugen 2007:190). This resulted in an agreement between the MPE and Statoil, stating that a CO_2 capture pilot should first be constructed: the Technology Company Mongstad (TCM) (MPE 2006b). The actual CCS requirement was more ambiguously formulated. The agreement merely noted that 'the parties have as a common objective to create a CO_2 solution within 2014 with concern for common industry practice for safe and rational execution of such projects.'

As we will see, this wording was indeed open to interpretation. Initially, CCS proponents expressed deep disappointment with this agreement, but later a unanimous Storting endorsed it in a show of political unity unprecedented in the history of Norwegian gas power (Innst. S. nr. 205 2006–2007; St. prp. nr. 49 2006–2007; Tjernshaugen 2007:192).

Interviewees have indicated that it was a challenge for the government to collaborate with the petroleum administration in negotiating the Mongstad agreement. Firstly, the politicians lacked technological expertise, and Statoil had a clear information advantage as regards Mongstad. Governmental expertise was spread among many agencies and subsidiaries, and the MPE itself was rather distanced from dealing with industrial decisions of this character (Boasson 2005; NVE 2006). Whereas the ministry had certainly had such expertise back in the 1970s, industrial investment decisions were no longer part of its daily routine. A former political appointee characterized the ministry thusly: 'the most severe problem was not that they opposed me, but rather that the ministry lacked competence'. Thus, the politicians had to rely largely on technical information from the environmental organizations (see Gjerset 2007; Jakobsen et al. 2005).

In the aftermath of the Mongstad negotiations, the government radically upgraded its expertise on CCS. Firstly it was decided that Gassnova, transformed into a state enterprise, should manage all kinds of state engagement in CCS prescribed in the governmental declaration (MPE 2008c; 2009a). Moreover, in autumn 2007, the ministry was re-organized and all CCS-related issues were gathered in a separate division for Climate, Industry and Technology. These organizational changes significantly enhanced the governments' in-house CCS expertise. Moreover, CCS competence increased outside the government as well. Most significantly, the dominant Norwegian petroleum equipment producers Aker and Kværner showed increasing interest in CCS, eventually merging to create the daughter company Aker Clean Carbon—Norway's sole provider of CCS technology (Aker Clean Carbon 2010; St. meld. nr. 15 2001–2002:44).

In 2005, CCS was no longer a strictly Norwegian issue. That year represents a turning point in EU climate strategy (Boasson and Wettestad 2013; Commission 2005b). At this stage, many European actors were intensifying their search for climate solutions and CCS attracted increasing attention from the European Commission as well some EU member states, particularly Germany and UK (Commission 2007a). Further, all the large European oil corporations were now engaged in CCS (see BP 2009; ENI 2009; Shell 2009; Total 2009). In describing the shift, a BP spokesman noted that ten years previously 'there were just a few lonely voices out there trying to make a case for carbon dioxide capture and storage (CCS). Now we have a full CCS choir singing from the same song sheet' (BP 2009). The Commission appointed a CCS technology platform with researchers, experts from the petroleum industry as well as pro-CCS activists like Bellona (see Commission 2005b:11).

Even though CCS was opposed by environmental groups as well as some governments, the EU adopted a directive on 'the geological storage of carbon dioxide' in 2008, providing a legal framework that would allow CO_2 to be stored safely (Boasson and Wettestad 2013). It also decided to earmark the profits from 300 million allowances under the EU Emissions Trading Scheme as support to CCS pilot projects and renewable energy investments. Unlike their Norwegian counterparts, many European environmental organizations worried that the stored CO_2 would leak, and that funding of CCS would tap money from renewable energy.

On the backdrop of rising international interest in CCS, Prime Minister Jens Stoltenberg changed his attitude towards CCS. Interviewees with leading administrative and political positions who have had considerable contact with Stoltenberg highlight his important role and extraordinary engagement. They also note that Stoltenberg supported the logic of societal cost minimization, not the technology development logic. He had initially opposed CCS, but in order to create a government he had to yield. In 2007 the future for Norwegian CCS development looked rather promising. In his New Year's speech to the Norwegian people, the prime minister stated:

> Our vision is that we, within 7 years, shall develop technology that enables us to cut carbon emissions. This will be an important breakthrough for mitigation of emissions in Norway, and I gather that the world will follow when we succeed. This is a large project. It is our moon landing.
> (Stoltenberg 2007, my translation)

Stoltenberg compared the challenges of constructing full-scale CCS to the US *Apollo* programmes of the late 1960s/early 1970s that enabled the first human to set foot on the Moon. The speech and the 'moon landing' metaphor brought widespread acclamation from the opposition: a unified political front for strong political engagement in CCS had emerged. However, the new CCS consensus conflicted with the minimizing societal cost logic that still dominated Norway's official climate policy at the rhetorical level. In 2008, all parties in the Storting, except the conservative-populist Progress Party, joined in a common climate policy agreement stating that:

> [G]eneral measures are the central feature of the national climate policy. Economic measures that cover all sectors create the foundation for decentralized, cost-efficient and informed action, wherein the polluter pays.
> (Innst. S. nr. 145 2007–2008:14, my translation)

In contrast to this rhetoric, the agreement included a wide range of measures pitched to ensure domestic action and technology development, for instance with respect to CCS (Innst. S. nr. 145 2007–2008:15; St. meld. nr. 34 2006–2007:15). Stoltenberg, traditionally aligned to the minimizing societal cost logic, now gave in to the technology-development character of the CCS policy. Also the Storting in full submitted to this approach (see for instance Innst. S. nr. 206 2008–2009; St. prp. nr. 67 2008–2009). Thus we may conclude that the broad agreement on climate policy was a true political compromise that glossed over intrinsic political discrepancies. Now political CCS competition primarily boiled down to quarrels over which party was the most true to the logic of technology development. Internally within the government, the very high state investments now designated for CCS served to heighten the importance of the Ministry of Finance. Due to the magnitude of the investments and the direct engagement of the state, CCS was continuously discussed in the 'sub-committee' consisting of Prime Minister Jens Stoltenberg and the leaders of the two other political parties in the Red/Green government coalition.

Despite the political harmony, the government soon encountered new CCS problems, relating to the actual construction of the two facilities but also regarding EU policy. Let us first explore the EU-related problems. Even though the EU paid greater attention to CCS as a potential climate solution, it had no history of accepting that state aid should be granted to this technology. As a main rule, the EU prohibits all state measures that distort or threaten to distort competition, that are granted by the state or through state resources, or when the intervention is likely to affect trade among member states (Community Guidelines 2001, 2008). Government investment in low-carbon technologies is regarded as state aid when the investment yields a lower rate of return than what would be acceptable to a private investor. EU guidelines for state aid specify that only the additional costs of a low-carbon power plant, compared to a conventional power plant, are eligible for support. In the end it is the technological quality of the project that determines whether it is eligible for support, not the costs. Further, state aid is legitimate only if all kinds of actors who adopt the same technical criteria receive the same level of support. It is the Commission, or EFTA Surveillance Authority (ESA) in the case of Norway, and not the member states, that has prime competence in this issue-area (Community Guidelines 2001, 2008). The ESA is to approve all state aid measures, and no measures shall be implemented prior to such endorsement. Should aid be granted illegitimately, the ESA may decide that the receiver must repay it.

Problems soon mounted after Norway's Ministry of Petroleum and Energy issued the Technology Centre Mongstad (TCM) for notification in accordance with EU regulations on state aid. Initially, the administration in the ministry argued that state aid to CCS should was merely a commercial strategy on the part of the Norwegian government and did not really constitute state aid (MPE 2007c:3). Because the government did not expect profits, the ESA did not accept these arguments (ESA 2007a; 2007b). Evidently, the ministry's line of argument did not fit the actual requirements in the EU regulations. The ESA repeatedly advised Norway to change its argumentation, but the ministry continued along the same lines.

The process entered a new phase after an internal swap in the Jens Stoltenberg II government, when Åslaug Haga replaced Odd Roger Enoksen as Minister of Petroleum and Energy. When the new political leadership took control over the processes, the arguments changed radically (see MPE 2007d; 2007e). The new political leadership aligned their reasoning to a technology-development logic, which also fit rather nicely with the intrinsic logic of the EU guidelines for state aid. Now the argument went that TCM would facilitate CCS technology development of European importance and that '[t]he Norwegian policy objectives are to a large extent similar to the policy objectives of the European Union' (MPE 2008b:7, 13).

Interviews show that it took considerable effort to get discussions moving. For half a year the ministry and the ESA met on a weekly basis and the Norwegian political appointees met with several Commissioners. The minister and her deputy minister also had several trips to the EU, where they argued for state aid acceptance. By making the ESA aware of the changes in the wider European environment in relation to CCS, and by activating commissioners and getting them to them intervene in the process, the political executives made the ESA attentive to the Norwegian arguments. Finally, the ESA agreed that state aid could be granted to the Test Centre Mongstad in July 2008 (ESA 2008c:22). Interviewees agree that this political intervention occurred at a critical moment and that the political steering was a precondition for final success. Later, the ESA agreed to state aid at Kårstø without further ado (ESA 2009). Uncertainty still remains as to how the ESA will deal with future notifications of the full-scale CCS at Mongstad, which will entail more state aid than TCM or Kårstø.

While state-aid discussions with the EU were challenging, the most severe blow to realization of Norway's CCS policy came from the industry. Calculations showed that installing CCS at the Kårstø plant would cost considerably more than anticipated, and due to unexpectedly high gas prices and low electricity prices the plant was not operating full-time (Gassco 2009:6: NVE 2006). Installing a costly CCS facility at a plant that was not in full operation seemed absurd. Thus the project was postponed and the government began to assess the idea of integrating capture of the carbon from the gas processing plant in the gas-powered CCS facility (Gassco 2009; Gassco and Gassnova 2010). By the end of 2010, no final decision had been made (Riksrevisjonen 2010b:151). Interviews have indicated that the operator of Kårstø, Naturkraft, was willing to install CCS, but only on condition that the state would cover all expenses. At this stage the problems were not mainly related to unwillingness, but to the simple facts that the gas-power plant had already been constructed without CCS, that the conventional plant proved to have low profitability, and that the highly complex industrial activities at the gas processing site where the gas-power plant was situated made CCS technologically challenging.

Also the Mongstad project faced difficult hurdles. In 2009 Statoil launched a master plan for capture of CO_2 from the gas-power plant and the refinery, estimating the CCS costs to be four times higher than indicated two years earlier (SFT 2006; StatoilHydro 2009:8, 10). Statoil also called for the CCS installation to be postponed with five years. This reflects both technological challenges, and the fact that conventional gas power had never been a pet project for Statoil, which was even less attracted to CCS. In order to understand this reluctance from Statoil—which, after all, has the state as its dominant owner—we must take into account the structural composition of the Norwegian petroleum industry.

Internal conflicts in the field or petroleum can help to explain the problems that mounted in relation to the Mongstad full-scale CCS plant. There is a clear demarcation line between offshore petroleum exploration on the one hand and onshore refineries and petrochemical industry (downstream activity) on the other. It was the latter actors that promoted construction of a gas power at Mongstad in the first place. Statoil encompasses both activities, but there has always been an internal organizational divide between the two (Statoil 2010). There is also a significant regulatory divide: offshore activities are regulated by the Petroleum Directorate and the Petroleum Law, whereas onshore activities regulated by the Energy Directorate and the Energy Law (Energiloven 1990; Petroleumsloven 1996). Several interviewees underline that because offshore activities are far more profitable than downstream activities, offshore actors clearly have the upper hand in Statoil (see StatoilHydro 2009). It was this weak part of the industry—the onshore refineries and petrochemical industry—that favoured the construction of conventional gas power (Tjernshaugen 2007). Only after more than a decade of urging from the onshore divisions did the Statoil executives agree to the conventional gas-power plant at Mongstad. The CCS component increased the cost, and was thus not very popular among the Statoil leadership or the Mongstad project organization.

The technology development logic from the former era of Norwegian petroleum activity has remained present in the downstream-related parts of Statoil and in the R&D department; after all, all kinds of petroleum activity are underpinned by advanced engineering. Certainly, a substantial number of people within Statoil developed a true involvement in CCS, particularly those responsible for R&D activities. For instance, one interviewee enthusiastically underlined: 'We really want to do this. (. . .) This is not only for show. We are really interested.' All the same, it is still the interests of the offshore divisions that determine the areas in which Statoil is willing to take financial risks, and thus far CCS has not been among these priorities.

As of the end of 2010, the future of the full-scale projects at Mongstad and Kårstø was still rife with uncertainty. The Mongstad TCM proceeded as planned, but this is only a technological test centre and does not ensure any actual storage of CO_2. Despite broad and strong political CCS support, the government had a hard time delivering on CCS, or, in the prime minister's own words, in achieving a 'Norwegian moon landing'. This grandiose promise gave the opposition parties a golden chance to attack the government, even though they agreed with its policy approach.

It is still unclear whether the politicians will emerge as the true victors in the conflict over full-scale CCS in Norway. In the following, we will first present a multi-field assessment focusing on the social forces that have shaped the policy outcome. Second, the significant entrepreneurial efforts involved in this process will be assessed.

4.4 ASSESSMENT

Process tracing shows that Norway developed a CCS policy because the politicians needed a compromise in the gas-power struggle. The Governmental Industry Development character of the policy reflects the technology development logic of those who originally launched the idea and opposition from the strong organizational field of petroleum made the policy very volatile, leading to steady increases in funding needed to realize a full-scale project.

Table 4.1 presents all the mechanisms in operation in the case. In the following we will discuss the relative importance of the three social fields for explaining why the Norwegian CCS policy emerged and why it had a Governmental Industry Development character in 2010. Also the relationship between social and entrepreneurial mechanisms will be examined.

Table 4.1 Mechanisms in the case of CCS

Phase ⇨ Social system ⇩	Phase 1, 1995–2000	Phase 2, 2000–2005	Phase 3, 2005–2010
Political field	*Politicizing*	*Legislature Governing* Importer Spider	*Legislature Governing* Importer Fashion Queen Shrewd Lawyer
Organizational field	*Segmented* Spider	*Segmented* Importer Fashion Queen Spider	*Segmented* Importer Spider
European environment	*1000 Flowers*	*1000 Flowers*	*1000 Flowers* *Unpredictable EU governing*

The Political Field: Strong Social Mechanism and Much Entrepreneurship

In the 1995–2000 period, the political situation in relation to gas power was unstable, hazy and fraught with conflicts. Gas power became a symbolic issue for Norwegian climate debate, but this debate was far from salient in the political landscape. It was not a main point for any of the political parties: their identities relied on other issues. Eventually, the politicizing of gas power also led to the politicizing of CCS. From being an 'off-beat' proposal on the fringes of the climate-policy debate, it became the focus of attention.

Gas power emerged as a watershed issue only *after* the minority coalition government had resigned as a result of gas-power disagreement in 2000. In the aftermath, the politicians came to need a compromise that could enable gas-power promoters and opponents to work together without losing face. But—supporters of gas power were in fact in the majority in the Storting throughout the whole period. Why then was a compromise needed? The explanation lies in the fact that the three political parties that promoted gas power—Labour, the Conservatives and the Progress Party—disagreed so deeply on other issues that a ruling coalition of those three was unthinkable. Due to their size, at least one of them would have to contribute in a government for there to be a sufficient number of parliamentary votes. In other words, one of the large parties would have to collaborate with the smaller gas power opponent parties: compromise had become necessary.

CCS was initially promoted by actors that had taken part on both sides in the gas-power conflict: the environmental foundation Bellona on the one hand and the trade unions on the other. This made it a tempting compromise candidate. In fact, CCS was also the only compromise candidate for which neither side of the gas-power conflict would have to admit that they had been overruled. Note that it was only *after* the politicians had chosen CCS as the compromise that it gained wider acceptance within the environmental movement, and an actual Norwegian provider of CCS technology emerged, in the form of Aker Clean Carbon.

By 2005, CCS had become an issue in which all the political parties had a stake: all, with the exception of the Progress Party, had been part of a government that supported this climate measure. Moreover, towards the end of the first decade of the 2000s, even the Progress Party—which officially doubted that climate change was caused by human beings—supported the CCS measures. In the course of only ten years, CCS, once regarded as a strange and unrealistic proposal, had become the order of the day; it had shifted status from heresy to dogma (see Hoffman [1997] 2001). In the run-up to year 2000, the social mechanisms of the political field shifted from Politicizing into Legislature Governing. The length of the period in which CCS was driven by the Legislature Governing mechanisms is remarkable: despite some ups and downs, CCS was largely subjected to political competition for ten years.

To what extent and how did political entrepreneurship contribute to this development? The combination of opposition from a segmented field and a strong political field led politicians to execute a remarkable amount of entrepreneurship: from lower-ranking political advisors and MPs, to cabinet ministers (particularly Einar Steensnæs and later Åslaug Haga), and the prime minister himself, Labour Party leader Jens Stoltenberg. This involved both institutional and structural entrepreneurship: one string of activities related to promoting CCS as a policy solution, and the second to the negotiation and translation of EU steering signals.

In 2001, the new government needed to make the CCS compromise look credible, but they distrusted the governmental administration. This led Einar Steensnæs to execute important Spider entrepreneurship, taking the highly unusual step of hiring a senior employee in Bellona. The close contact with the environmental movement gave politicians access to significant counter-expertise. In the 1995–2000 period, environmentalists had approached politicians aggressively, but eventually the politicians became the active part, constantly searching for information that could be used against their own governmental apparatus. Such information enabled politicians to take decisions that would not otherwise have been possible. And, as the environmental groups were carriers of the old institutional logic of the petroleum field, the technology development logic, they exposed politicians to this way of thinking—which eventually put its mark on the CCS policy outcome.

In 1999, the government decided to strengthen Norwegian pollution regulations, without making any reference to Europe. In 2000 the Storting opposed this, with Jens Stoltenberg in the leader seat, referring to the lack of such regulatory practices in Europe. This was apparently effective, because it contributed to enable the construction of conventional gas power in Norway. In the long run, however, this interpretation came to appear as non-legitimate. Jens Stoltenberg and other pro-gas-power parliamentarians exerted Importer entrepreneurship that later was overruled by the Importer entrepreneurship of anti-gas-power politicians. The question was whether the dominant standards within the EU actually warranted Norway's strengthening its practice, or if it could continue to allow the construction of carbon-emitting facilities.

There are two reasons for the change of perceptions. First, the outgoing coalition government (Christian Democrats, Centre, Liberals) interpreted these arguments as 'an attack on the Norwegian Pollution Control Act'. This creative rhetorical move was to have significant long-term effects. Even though the features of the original emissions permit had actually been far stricter than conventional pollution practice, the first emissions permit came to be understood as the appropriate way of executing the Pollution Act. Second, MPs and lower-ranking political advisors who opposed conventional gas power repeatedly referred to EU pollution regulations in arguing that Norway should continue to regulate carbon by its Pollution Control Act. They continued to do so even after the EU had signalled that carbon should be regulated by the emissions trading system rather than national pollution legislation. This move proved highly successful; after 2005 it became impossible for supporters of conventional gas power to employ 'European' arguments to legitimize the development of conventional gas power. Skilled Importer entrepreneurship ensured that the effect of 'Europe' shifted, from being an obstacle to an enabler of Governmental Industry Development

CCS policy. As we have seen, this creative interpretation of EU policy proved to be a necessary precondition for the later ban on gas power without CCS.

Paradoxically, Jens Stoltenberg eventually shifted from performing anti-CCS entrepreneur to a pro-CCS position. He had initially argued that Norway should not adopt stricter carbon pollution regulations than other EU and EEA countries. Later, however, he argued that Norway should promote CCS actively, because this was a European climate solution, perhaps even a global one, and that Norway should be a frontrunner. Stoltenberg was not the only actor who served up such Fashion Queen arguments, but he provided the boldest example of this argumentation in his 'moon landing' metaphor. CCS had become a European 'fashion' and Norwegian politicians aimed to be the leader of the trend: the rational was that 'Norway does the same as all the other countries, only it does it better.' Note that Stoltenberg also turned from being the chief political defender of cost-minimizing logic as the dominant approach in Norwegian climate policy, and became a main defender of strong direct state engagement in CCS.

Due to the structural position of Jens Stoltenberg, his 'moon landing' metaphor helped to transform CCS from a much-needed political compromise to a central objective in its own right; he had promised that Norway would be at the forefront of European developments, and now his government had to demonstrate that they really were. This proved far more difficult than anticipated. Still, we may conclude that, by playing the fashion game, these entrepreneurs ensured that European developments strengthened the governmental engagement in developing CCS in Norway. Great expectations were created by this argument. Because the government risked losing political credibility if the CCS endeavour were to fail, they had to continue along this path. The politicians had played 'the fashion card' in a way that restricted their own leeway underpinning the Legislator Government mechanism and making it harder for them to back down once it became evident that ensuring the construction of a full-scale plant would be no easy matter.

Åslaug Haga, Minister of Petroleum and Energy in 2007 and 2008, engaged heavily in the negotiations over EU state aid notification, executing Shrewd Lawyer entrepreneurship. The aim was to ensure that generous funding and the Governmental Industry Development approach was accepted. Even though the EU has substantial authority as regards state aid, it did not have a firm hand on the tiller in the CCS state-aid case in Norway. National responses to European involvement largely determined the final outcome. At outset there had been considerable misfit between the perceptions of the MPE and ESA. The initial clumsy approach of the ministry created severe problems for realization of the government's CCS ambitions. In order to create a Governmental Industry Development policy on CCS, this obstacle had to be overcome. It seems safe to say that political entrepreneurship was crucial in ensuring the final outcome in relation to TCM. Through active and creative engagement, the Norwegian political executives succeeded in creating a formal ESA decision that underpinned the development of Norwegian CCS policy: skilled entrepreneurship ensured that the effect of Europe influence shifted, from being an obstacle to an enabler in the case of CCS.

We may conclude that both Legislature Governing mechanism and the ensuing volume of entrepreneurial acts helped to ensure that the political field had a strong imprint on the policy outcome. Political entrepreneurs were to a high degree able to circumvent, undermine and alter formal distribution of authority by shifting the information basis on which formal policy CCS decisions were made. They also significantly changed how EU regulations were perceived. Even though the policy outcome was based on input from actors outside the political field—CCS researchers, trade unions and the environmental movement—it genuinely reflected the functioning of the political field: at no point in the ten-year process did the political field merely adopt arguments provided by the segmented organizational field. But when it came to ensuring stability around the policy area, and ensuring actual construction of a full-scale CCS facility, the politicians seemed rather powerless. No matter how strong their commitment, and how clear their signals towards the industry became, the problems persisted.

Political field mechanisms do not give the full explanation of this enduring dominance of the strong Legislature Governing mechanism. Let us now explore how resistance from the organizational field forced politicians to take the lead in policymaking, year after year.

The Organizational Field: Providing Repercussions, but also Seeds of Change

The character of the organizational field is strikingly similar to the classic, theory-oriented descriptions of segments, iron triangles and policy monopolies (see Baumgartner and Jones [1993] 2009; Egeberg et al. 1978; Hernes 1983:290; Lieberson 1971; Lowi 1969). In the case of Norway, commercial and governmental actors were closely related structurally, through economic interdependencies, regulatory ties, and governmental ownership structures. Not only did Statoil take the investment decision on CCS on Mongstad, it also initially controlled the technological information related to CCS. The governmental organizations have seldom challenged the commercial actors; instead, they have collaborated closely. Further, for most of the period, government agencies strongly resisted CCS initiatives from politicians. From 1995 and onwards, the centralization of structural resources was strengthened; the privatization of Statoil and the 2007 merger between Statoil and Hydro were important in this respect.

The field is clearly dominated by one professional logic: commercial actors and regulator alike adhere to the market logic. This logic, strong already in the early 1990s, gained increasing hold in the course of the decade. This goes for the commercial organizations as well as the regulators. In most respects, the character of the policy on CCS clashes with this logic. The minority actors that promoted CCS within the field—technological researchers, but also the trade unions—were still basing their reasoning on the technology development logic that had dominated the whole field in the 1970s and 1980s. Now this logic re-emerged in the political field, gaining wide and strong political backing—but the organizational field remained hostile. One reason was that the logic of governmental industry development clashed with the market logic: technological risks could entail unforeseen costs that might jeopardize the market value of Statoil. In turn, that would reduce the revenues of the largest shareholder: the state, represented by the Ministry of Petroleum and Energy.

The market logic provides business organizations with the main initiative: the government is expected to provide general and overarching economic measures, while the business organizations select the actual projects. Moreover, it turns the commercial actors' interest towards the most profitable investments. However, CCS policy discussions in Norway have been underpinned by the premise that realization of CCS should lead to reduced emissions at two specific plants (Kårstø and Mongstad), and not ensure that the potentially most profitable CCS projects are constructed.

The reluctance of the organizational field had the paradoxical effect of strengthening the technology development character of the CCS policy outcome. Two different factors related to institutional logic drove this counter-intuitive effect. First, the industry's interest in gas power was rather lukewarm at the outset; conventional gas power was favoured primarily by marginalized actors within onshore branches of the industry. To be sure, the industry became more interested when it seemed that gas power would be profitable, but then, when the expected profitability dropped, interest evaporated. Not surprisingly, adding a costly CCS facility to the Mongstad plant which already had low profitability was not tempting for Statoil.

Thus, the government found itself in a very bad negotiating position in 2005, when it tried to persuade Statoil to commit to building a gas-power plant with a CCS facility. This led to an agreement between the two that merely obliged Statoil to *aim for* installation of CCS, resulting in the actual construction of a gas-power plant but later postponement of the CCS facility. Moreover, the state had pledged to cover all costs, providing no incentives for Statoil to minimize expenses. And so, the state was later faced with record-high demands for state aid. Because neither gas-power development as such nor the CCS technology seemed to offer any immediate market advantage, interest remained lukewarm. True, corporate actors may have appreciated the positive effect that CCS engagement could offer to their image, but this was not enough to make them more committed.

Second, the initial reluctance to act on CCS signals from governmental organizations forced the government to develop its own CCS expertise. Together with government agencies, the petroleum corporations actively refuted the claims from political executives. When the CCS endeavour became specifically directed at Kårstø and Mongstad, the need for technical expertise within the government grew even stronger. Because these were complex industrial sites, the information barriers were formidable. Statoil was a dominant owner of the Kårstø and Mongstad plants, and the government could not possibly check every piece of information given concerning why it would be so cumbersome to install a CCS facility. Further, the agreement between Statoil and the government ensured Statoil information control. Statoil was given responsibility for the preparation of the CCS facility—but not for the costs it entailed.

By strengthening Gassnova and boosting expertise within the ministry, the government tried to counter the information advantage of the commercial actors. While these measures eventually empowered the government, this also led the state to become increasingly committed to the logic of technology development, and so the Governmental Industry Development character of the policy grew stronger over time. This was not popular with the industry. On this backdrop we may conclude that the effects of the segmentation mechanism proved somewhat different than those indicated in classical political science contributions. Because of heavy reluctance, the policy outcome gradually came to contrast more and more with the preferences of the field actors, rather than the other way around.

The segmentation mechanism played a key role in the CCS story, but the field also contained the seeds of change. We see much entrepreneurship from marginalized organizational field actors. It was researchers from within the field who initially come up with the CCS idea, approaching environmental organizations that in turn approached politicians. The CCS researchers and the environmental organizations performed important Spider entrepreneurship. Initially, only Bellona and a few petroleum researchers had championed CCS, but the idea appeared more realistic once technological solutions had matured within industrial R&D groups. Later the environmental foundations Bellona and ZERO used networking to ensure couplings between people with technical expertise and political executives. Bellona and ZERO operated at the boundaries of the field, with significant links to field-level actors, but also close ties to central politicians. To some extent, the development of counter-expertise within the environmental foundations helped to even out the petroleum actors' information advantages.

The environmental organizations also applied institutional entrepreneurial techniques to promote CCS as a climate solution. Throughout the decade, Bellona and ZERO argued that if the Norwegian government ensured full-scale CCS development, Norway would be leading the way in the international efforts at carbon mitigation. Initially, this argument lacked clout because CCS was of little interest in Europe, but this line of argumentation became effective once CCS started to attract some attention around 2005. However, it was only when important politicians began employing European arguments to legitimate their CCS offensive that Fashion Queen entrepreneurship gained clout, helping to underpin the Legislator Government mechanism of the political field.

Assessment of the organizational field mechanism shows that, despite the segmented nature of the field, it contained the seeds of change. The field-level CCS promoters succeeded because they applied multi-field entrepreneurial techniques: by using the environmental organizations as mediators they were able to circumvent the dominant organizational field-level actors, and plant the idea in the political field. Continued resistance from the powerful field forced the politicians to go out of their way in order to try to effectuate their CCS ambitions. The entrepreneurial activities of petroleum researchers were important in the early phases of policy development: later, other actors took over. Due to the segmented state of the field, there was no need for powerful actors like Statoil and the administration of the Ministry of Petroleum and Energy to deploy entrepreneurial techniques in order to impede realization of full-scale CCS.

European Environment: Small Direct Effect on the Norwegian Policy Process

The European Environment initially had a 1000 Flowers character. CCS emerged on the Norwegian political scene when CCS had hardly been debated elsewhere in Europe. No other European countries had developed CCS policies and the EU had showed scant interest. The EU regulations on state aid applied to this area, simply because they concerned all areas in which state aid was offered, but no specific state-aid guidelines in relation to CCS existed. Moreover, the pollution regulation regimes in Europe had not yet become attuned to the new major environmental issue on the political agenda: climate change. Thus we can safely conclude that Europe had scant impact on the emergence of this string of policy. Here we should also note that there was nothing within the European field that actually obstructed the development of a national CCS policy at this stage, even though some Norwegian actors tried to give this impression.

The pollution regulation of the EU as well has rather strong technology-development characteristics: it is the national governments that are to set national pollution limits and thus to engage directly in selecting winner technologies. The development of Norwegian CCS policy was full of references to EU pollution regulations, but processes tracing shows no direct relationship between EU policy and the national policy outcome in Norway. EU policy was subjected to quite creative, entrepreneurial interpretations from the Norwegian actors. For instance, the EU explicitly exempted carbon from its pollution directive, IPPC, in 2003—but all the same, the Norwegian government referred to this directive in arguing that CO_2 from gas-power plants should be granted pollution licenses for carbon. This decision was a necessary precondition for a key decision that was taken later: that no gas-power plants without CCS facilities would be granted permission to operate.

The EU regulations on state aid have an unpredictable EU Governing character. As a main rule the EU allows its member states to grant state aid if the technology is of superior quality, irrespective of its actual cost efficiency, but EU state aid decision-making may change considerably from case to case—there is evidently significant room for negotiating. At one point it nonetheless seemed as if the EU regulations on state aid could pose a threat to the direct engagement of the Norwegian government in the CCS technology endeavour, but this issue was eventually resolved. Moreover, this conflict did not really centre on the actual approach chosen by the Norwegian government, but on the rationale used by the MPE in initially justifying its policy choices. Later, the Minister of Petroleum and Energy, Åslaug Haga, changed the line of argumentation and ensured endorsement for Norwegian state aid.

CCS was far more controversial elsewhere in Europe, but empirical mapping indicates that this scepticism never attracted much attention in Norway. European interest in CCS increased after 2005 and Norwegian CCS promoters referred to European developments in order to bolster their claims. Whereas European developments undoubtedly inspired Norwegian environmentalists and politicians, the positions of the commercial actors seem to have remained unaffected—so it is hard to argue that Europe had significant impact on the development of CCS policy in Norway. Note that the adoption of an EU CCS directive in 2008 had scant impact on Norwegian CCS discussions. This directive concerned technical and legal challenges relating to storage of CO_2, an issue that had never been a significant political challenge in Norway.

By creatively interpreting the situation in the European environment, Norwegian politicians were able to overcome resistance from the organizational field that initially hindered the emergence of a Governmental Industry Development policy. Hence, the European environment influenced the development of CCS policy primarily through entrepreneurship.

4.5 CONCLUSIONS

This chapter has examined and assessed why the Norwegian CCS policy emerged and why it had a Governmental Industry Development character in 2010. We have seen that this was possible because the political field gained a Legislature Governing character, and countered opposition from a strong, segmented organizational field. The CCS issue was so important for politicians that they repeatedly acted at entrepreneurs, trying to enhance their power over field-level processes—and partly succeeding. Moreover, we have seen that even the most segmented organizational fields can carry the seeds of change: the CCS idea originated from marginalized actors within the organizational field of petroleum

The political field had an independent role, although the powers of the politicians were clearly constrained by the segmented field. In some respects, the policy outcome may seem like a mountain giving birth to a mouse. Many politicians devoted massive amounts of time and energy to the issue, and the high-level political game was full of drama. And yet, no large-scale CCS facility has been constructed in Norway, and it is uncertain whether that will ever occur. But the policy outcome can also be seen in a more positive light. A ban on conventional gas power is now in place—and that was, after all, what the environmental movement originally fought for. Some conventional gas-power plants have been developed, but the many plants originally envisioned will not be realized. Moreover, the Norwegian government developed an enduring commitment, a specific organization (Gassnova) committed to promoting CCS, and the government is heavily engaged in R&D on CCS.

Political entrepreneurship played an important role, but we must not be so dazzled by this entrepreneurial achievement that we fail to recognize how the social mechanisms of the political field and the national organizational field played the most crucial parts in this drama. However, the fact that politicians played key entrepreneurial roles is a highly interesting finding, given the scant attention paid to politicians in the literature on entrepreneurship. The broad and strong support for CCS in the political field might have been expected to be a stabilizing factor; but evidently this was not enough to ensure stability. Political entrepreneurship had scant effect on the segmented nature of the organizational field. It was developments not directly related to CCS that served to ensure that the field became more segmented and resistant to political steering over time.

NOTE

1. The Norwegian Water Resources and Energy Directorate (NVE)

5   Entrepreneurship Paradoxes
Renewable Energy Policies

5.1 INTRODUCTION

The production of renewable stationary energy boomed in Europe after the turn of the millennium. Wind power was the most popular renewable technology by far, but renewable district heating attracted increasing attention in the latter half of the first decade. Also in Norway, new renewable energy sources gained in popularity. But in contrast to the dominant European pattern, district heating policy developed incrementally from 2000 to 2010, whereas the state aid scheme for wind power gave rise to many conflicts and considerable uncertainty (Riksrevisjonen 2010a). District heating plants mushroomed, while wind-power development stalled. This chapter explores the development of Norwegian renewable energy policy since its initiation in 2000 and up until 2010 in two sub-cases: renewable electricity and renewable heating.

In 2000, as much as 99 per cent of onshore stationary energy production in Norway was renewable, mainly from large hydropower plants (MPE 2008a). Thus the country's stationary energy production was practically without any carbon footprint at the outset. Because there was political agreement on preserving (i.e. not damming or regulating) the country's remaining major streams and waterfalls, and gas power gave rise to major political disputes, new renewable energy sources gained. Large-scale wind power seemed well-suited for Norway's heavily electricity-dependent energy system, whereas renewable heating was alien to Norwegian utilities and would require considerable development of new infrastructure. Hence, wind power gained wide support while there was scant interest in renewable heating. Moreover, many entrepreneurs became involved in supporting renewable electricity, whereas few engaged in policy discussions related to district heating. And yet, it was renewable electricity that came to draw the shortest straw.

The state policies promoting the two energy sources that were adopted in 2000 were strikingly similar, both consisted of state aid measures aimed at minimizing societal costs. While the district heating scheme remained uncontroversial, the electricity measure gave rise to major conflicts and considerable political controversy. In response to the protests, the governments of Norway and Sweden in 2010 finally agreed to launch a common market-based support scheme for renewable electricity: a green certificate market (MPE 2010c).

Norwegian renewable energy policies seem puzzling, in light of the renewables policies otherwise dominant in Europe. The 'fiscal incentives'–type district heating scheme is quite unique to Norway, most EU member states have adopted feed-in schemes, not certificate schemes for renewable electricity. *Why then did Norway adopt renewable heating and renewable electricity policies, and why had the former a fiscal incentives character in 2010, while the latter had a market character?* Here we explore two policy cases; even though the same players are crucial in both, the policy results are very different.

In terms of theory, of particular interest are the differences in entrepreneurship that played out in relation to the two sub-cases, with considerable activity targeting renewable electricity and very little directed at renewable heating. In-depth exploration of the two sub-cases enables us to assess why one case gave rise to so much entrepreneurship and not the other, and the relationship between entrepreneurial and social mechanisms over time. Whereas the literature on entrepreneurship has focused on how entrepreneurs may drive processes of change, these cases shows that change may happen without entrepreneurship and that entrepreneurs should be careful about what they wish for: despite good intentions, they may obstruct policy development more than facilitating it. Moreover, this case study can provide a good empirical backdrop for discussing how social conditions affect the space for entrepreneurship and how entrepreneurship may contribute to changing the functioning of social mechanisms.

5.2 POLICY OUTCOMES: FISCAL INCENTIVES AND MARKET MEASURES

The two policy outcomes for renewable energy in 2010 fall into two categories. The state aid scheme for renewable heating belongs in the category of fiscal incentives: here the state is directly involved in decision-making, determining which projects are to receive support and which are not. Projects are selected on the basis of economic, not technological, criteria. Also the green certificate scheme is based on economic criteria, but here the state governs indirectly by creating a market which in turn produces economic incentives.

In the period 2000 to 2010, two sub-targets existed: to produce 4 TwH renewable heating and 3 TwH wind power by the year 2010 (St. meld nr. 29 1998–1999). As of 2010, the government aimed at realizing 18 TwH new renewable energy, stemming from new heating or electricity production, or energy savings, from 2000 to 2011 (MPE 2010a). In addition, the Norwegian government was committed to financing 13.2 TwH in new renewable electricity production (including small hydro) by 2020, in the common Norwegian–Swedish certificate market (MPE 2010a).

The prime sources for district heating are waste heat from industrial processes, biomass and incineration of waste. District heating in Norway is in fact not 100 per cent renewable. Although the renewable share varies, this chapter refers to all district heating as 'renewable energy'.

The Norwegian state aid scheme for renewable heating is operated by the state enterprise Enova. Enova applies net present-value calculations and expected energy prices to calculate the support level; state aid is granted only to the applicants that require the lowest volumes of support in order to realize their projects (MPE 2010a). Each year from 2006 to 2009, some 40–55 million Euros were granted for district heating investment. Enova operates three sub-categories of district heating support: local heating construction, district heating construction and large-scale infrastructure development (Riksrevisjonen 2010a:39). Applications within each of the two latter sub-categories are compared internally, and support goes to the projects that will yield most energy production. State aid grants relating to the first category are determined by the application of technological criteria; all that fit the definition of 'local heating' may be supported (Riksrevisjonen 2010a:60). Thus, we see some elements of technology standards, but economic criteria dominate.

Also the wind-power programme in operation in 2010 was based on cost-efficiency criteria and investment support. Approximately 60 million Euros were granted for wind power in 2009, the first time in three years that any grants were given to wind power (Riksrevisjonen 2010a:32). However, this was replaced by a Swedish–Norwegian certificate scheme in January 2012 (MPE 2010c). The green certificate scheme involves a governmentally-induced market for renewable energy securities. Various governmental regulations determine the functioning of the market, with the key factor being the size of the quota that renewable energy producers are obliged to produce or purchase. What the purchaser of a green certificate buys is not the actual energy, but a security that confirms its economic contribution to the operation of green electricity somewhere within the area where the scheme applies. In contrast to the Enova scheme, such schemes yield operational support, not investment support; and the extra cost is paid by the consumers, not by state funds. Green certificate schemes will yield sizeable profits for companies that can produce renewable energy efficiently, and will favour actors large enough to manage considerable financial risks (Commission 2008a).

5.3 SILENT SUCCESS FOR DISTRICT HEATING, NOISY WIND POWER CONFLICTS

Phase 1, 2000–2005: Creating Cost-Efficient Support Regimes

Due to the abundance of hydropower, new renewable energy technologies attracted less attention in Norway in the 1970s and 1980s than in other

European countries. The government started to offer investment support for renewable electricity in the mid-1990s, but this support was small-scale and was granted *ad hoc* (Gjerløw 1996). Things changed around the turn of the millennium, when the Storting began to worry that the country might risk becoming dependent on imports of electricity. This must be understood against the backdrop of the new ban on major new hydropower development in Norway, which was supported by a large parliamentary majority (St. meld. nr 37, 2000–2001).

In 2000, the Storting voted to increase renewable heating by 4 TwH and wind power by 3 TwH by the year 2010 (St. meld nr. 29 1998–1999). This decision was taken on the same day that the minority coalition government resigned over the gas-power issue (see chapter 4). Despite scant attention, it represented a turning point in Norwegian renewable energy policy. The decision led to the creation of the state enterprise Enova, which was instructed to grant state aid in the most cost-efficient manner possible—which in turn meant giving priority to the diffusion of low-cost technology, rather than technology development (Ot. prp. nr. 35 2000–2001). This decision was underpinned by the cost-minimizing logic that dominated Norwegian climate policy at the time (see chapter 4). In parallel, a new principle was introduced in stationary energy policy: that high-quality electricity should not be applied for purposes where low-quality heating could suffice. This meant that electricity-based heating should be converted to district heating: a novel and radical challenge for Norway's heavily electricity-dependent energy system. Nonetheless, the principle was introduced without much controversy, and has remained basically uncontested ever since.

Two years later, the politicians increased the level of ambition regarding renewables somewhat by introducing a holistic 10 TwH target for energy efficiency and new renewable energy, in addition to the two sub-targets for wind and heating (MPE 2002). From this stage on, politicians did not distinguish between the two energy carriers when setting overarching political objectives: such priority-setting has been delegated to Enova (see MPE 2007a; MPE 2010a; St. meld. nr 34 2006–2007). In the ensuing years, district heating measures and renewable electricity measures were to embark on two highly different developmental paths. Whereas district heating developed in line with the decision taken in 2000 and attracted little attention, the EU, the Norwegian utilities and the environmental organizations challenged the Enova electricity support scheme, due also to the utilities' growing interest in wind power and their embeddedness in market logic.

Concerning district heating, there were some small, incremental developments after Enova was created and up until 2005. The absence of conflicts over heating was an indication not so much of wide support as of low interest. Regulators and commercial actors alike have traditionally had as a chief aim to ensure sufficient supplies of electricity, and predominantly large hydro (Midttun 1987; Thue 1996). Initially, only a handful of municipally-owned utilities were engaged in district heating activities; their instructions were to engage in combustion of municipal waste (see BKK 2001; Statkraft 2001). Because wind power represented an alternative technology for large-scale power production, it began to receive increasing attention, but heating was not deemed attractive; as one interviewee bluntly put it: 'frankly, we did not care much for heating then'. The position of the Norwegian commercial actors deviated from the case elsewhere in the Nordic countries: in Denmark, Sweden and Finland, renewable heating boomed in the early 2000s (Commission 2004). However, despite the lack of public debate or active support at the corporate level of the power companies, Enova and the power companies developed a good dialogue, and the companies prudently set about applying for support for a handful of projects.

Only months after deciding to create the societal cost minimizing Enova scheme, the majority in the Storting called for the Labour government to consider introducing a market measure: a green certificate scheme (Budsjett-innst. S. Nr. 9 2000–2001). Interviews show that this was a result of lobbying from Statkraft (the largest Norwegian utility) and environmental organizations. At this stage, few political parties had developed any firm views as to which support scheme for renewable energy they preferred: they merely wanted to voice the call for a more ambitious policy. However, the Storting's green certificate decision in 2000 led to subsequent parliamentary reports on the issue, all in turn exploited by green certificate supporters as opportunities for further promoting green certificates.

In 2001, a Liberal/Conservative minority coalition took office. Labour now joined the majority in the Storting and instructed the new government to embark on negotiations with Sweden on developing a common certificate scheme (Innst. S. 167. 2002–2003:17–18). These repeated instructions to the government must be understood in light of the minority government situation in Norway. None of the three governments in the period 2000 to 2005 had a parliamentary majority. The Storting had no clearly defined role in relation to renewable energy schemes. Nonetheless, renewable energy policy provided a good opportunity to criticize the party(ies) in government. As noted in chapter 4, Labour was strongly rooted in a cost-minimizing logic, and also the parties on the conservative side preferred climate measures of this kind. Interviews show that because the political salience of renewable energy was limited, few politicians actually delved into the details of the green certificate scheme. Most political energy was devoted to the gas-power struggle and the drama that unfolded when electricity prices peaked in winter 2002/2003.

In 2003–2004, lobby pressures for green certificates increased because the utilities united in a joint campaign for the development of an ambitious, technology-neutral, market-based green certificate scheme (MPE 2005a). This preference for a market scheme reflected the major shift in thinking that had taken place among the stationary energy producers in the late 1980s. Until then, the field had been embedded in the logic of technology development, and had focused on enhancing large hydropower production

(Midttun 1987; Thue 1996). The price of power was politically governed, and the companies were state-owned. The government estimated future energy demand, and power producers adjusted their production in line with this. By the late 1980s, however, it had become evident that the companies and the state alike were suffering economic losses because the price of electricity was too low to cover the costs of the new hydropower plants. Moreover, economists had gained central positions in the companies and in governmental organizations. And so the field now embarked on a profound liberalization reform.

Historians have described how the reform brought a shift from the logic of technology development to a market logic among Norwegian power producers (Thue 1996; Nilsen and Thue 2006). Interviews confirm that by 2000, the main objective of corporate actors had become to maximize profits rather than to ensure specific volumes of power supply. Introduction of the market logic led the companies to intensify their search for profitable new, alternative investment options. Because wind power was a large-scale renewable electricity technology, it now appeared particularly attractive to the commercial actors.

Within the governmental agencies, the situation was a different one. The energy divisions within the Ministry of Petroleum and Energy (MPE) and the Energy Directorate[1] shifted from reliance on technology development logic to the logic of minimizing societal costs. Previously, the Energy Directorate had encompassed the major power producer, the transmission system operator and a large construction advisory body (Thue 1996). Now these had been privatized, and the Energy Directorate was merely a regulatory body. The chief role of the governmental agencies became to ensure that the energy producers did not reap excessive profits. One interviewee explained: 'It is important to do societal cost calculations. This is the key to how the Energy and Water Resources Department in the ministry functions and thinks.' When Enova was established in 2001 this took place without much interference, the politicians let the MPE administration design the support schemes according to their preferences. Thus, the support measures for electricity as well as for heating were based on the cost-minimizing logic dominant in the ministry.

During the 1990s and early 2000s, the major Norwegian power producers underwent a phase of professionalization, marketization and growth. The green certificates fitted hand-in-glove with their new way of operating. A wave of mergers among municipally-owned companies ensured that the ten largest of Norway's 200 energy producers controlled 70 per cent of the country's power production by 2008 (MPE 2009c). Statkraft was the former Norwegian state power authority. In the early 2000s it represented 30 per cent of production capacity and had in addition substantial minority ownership in a handful of other large power producers. Moreover, it had taken over power plants in several European countries (Nilsen and Thue 2006). Because Statkraft had also engaged in mergers, it remained the largest Norwegian utility by far, the leading actor, commanding superior economic and analytical resources.

Eventually, most corporate actors, including fully state-owned utilities and grid operators, joined together in the Norwegian Electricity Industry Association. Coordination of lobbying efforts became a key task of this association, where Statkraft became the dominant actor.[2] Structurally, the field was dominated by the Ministry of Petroleum and Energy on the one hand and Statkraft on the other. Due the state ownership of Statkraft, the two were rather tightly coupled. Both focused on economic criteria, but the former was embedded in the logic of minimizing societal costs, while the latter was embedded in market logic.

Norway's major hydropower producers became involved in the European power-producers' promotion of a common EU green certificate scheme early on. Moreover, the first Norwegian wind-power constructions, in the early 1990s, were financed through Dutch green certificates (Riksrevisjonen 2010a:48). In the early 2000s, Statkraft actively referred to the development of an EU renewables directive in order to strengthen its own green certificate campaign. The argument went that Norway had to keep up with European developments and that the EU was now about to develop a green certificate scheme (see e.g. St. meld nr. 37 2000–2001). Eventually, the utilities' campaign for green certificates led to negotiations between Norway and Sweden on a common scheme.

How did the Norwegian green certificate discussion relate to European renewable energy developments? As in Norway, the large European utilities had shifted from a technology development logic to a market logic, and they started to promote the development of an EU green certificate scheme already in the late 1990s (Boasson and Wettestad 2013; Foquet and Johansson 2008; Rowlands 2005). In this they were strongly supported by European Commission officials. Concern was expressed that renewable energy might be introduced where the support schemes were the most generous, and not where the most efficient production could be ensured. Small-scale wind power in areas with little wind is an example of what they wanted to avoid. Unlike Norway, some EU countries have significant renewable energy industries, and they cried out against the green certificate offensive. They are a diverse group of enterprises, specialized in various different technologies, but all generally rooted in the logic of technological development (see EREC 2008). Most of these companies have emerged as a result of governmental feed-in support schemes. Feed-in schemes provide reliable support at a fixed level; and the more costly the technology, the higher the governmental support tends to be (Commission 2008a). Prominent EU member states with feed-in schemes, Germany in particular, mobilized to defend technology-development approaches and obstruct the introduction of a pan-European green certificate scheme.

The renewable actors prevailed, and the EU Renewable Energy Directive, finalized in 2001, came to permit a range of national schemes (Directive

2001). It referred to market instruments as favourable, but had no binding requirements in this regard. The EU Directive introduced a non-binding target of 22 per cent renewable electricity by 2010, with no commitments as to developing a market-reliant support scheme: member states were free to choose what support scheme to rely on (Directive 2001). Subsequently, member states continued to refine their renewable energy support schemes in line with their own national traditions. The Dutch scheme that had financed several Norwegian wind-power projects was altered, and thus Norwegian projects were held ransom (Riksrevisjonen 2010a:48). Sweden was one of the few countries that actually introduced a green certificate scheme (Energimyndigheten 2005), which began operating in 2003.

This European development obviously weakened the demands of green certificate supporters in Norway. However, wisely refraining from highlighting the retreat of the European green certificate plans, they continued to frame green certificate schemes as the most prosperous scheme and called for Norwegian governments to implement the renewables directive (see MPE 2005a). Moreover, they referred to the high support levels in the Swedish green certificate scheme, arguing for the same in Norway (MPE 2007b). This was smooth sailing: Enova was still in the process of establishing its fiscal incentive scheme and was in no position to defend the scheme publicly, few Norwegian actors were informed about recent developments in Europe, and no significant Norwegian actors were promoting the development of a Norwegian feed-in scheme. Moreover, the parliamentary opposition was easy to persuade because they wanted to portray themselves as being more on the offensive than the government—but, as we will see below, they were not strongly committed to the market logic on which the green certificate idea was based.

However, the EU also affected the discussion on renewable electricity policy through its rules concerning state aid. In general, EU regulations on state aid prohibit public measures that distort or threaten to distort competition, that are granted by the state or through state resources, or when the intervention is likely to affect trade among member states (Community Guidelines 2001). The Community Guidelines on environmental state aid specify the conditions under which renewable energy support may be provided. Investment support is not to exceed 40 per cent, although up to 100 per cent of eligible costs may be granted. Eligible costs are calculated by an extra-cost approach, which implies support corresponding to the additional costs of the renewable energy plant compared to a conventional plant. Due to the high costs of biopower development, EU exempted support to all kinds of energy production from biomass from the regulations. The rules do not restrict support by nation-states for energy deriving from biomass, so only support to electricity support, and not to district heating, is affected by the EU regulations on state aid.

Norway is required to notify the EFTA Surveillance Agency (ESA) of all state aid measures (as member states do to the Commission): no schemes are to be implemented without such endorsement (Community Guidelines 2001, 2008). If aid has been granted in a manner deemed not in compliance with regulations, the recipient must repay it—a rather severe compliance mechanism. There is no trace of societal cost minimizing in the regulations on state aid. On the contrary: even though feed-in and certificate schemes often represent extra societal costs and these schemes seldom include mechanisms to ensure calculation of the cost–value fraction, they are exempted from the regulations. Specifically, such schemes are exempted from the notification requirement if they are financed by consumers and not by the state (Kuhn 2001).

The Ministry of Petroleum and Energy has formal responsibility for ensuring that Norwegian renewable state aid is developed in line with these EU regulations. In 2002 the ESA called for notification of Norway's support scheme for renewable energy. Because biomass was exempted from the regulations, the Norwegian renewable heating scheme was not discussed. Correspondence shows that the ministry did not take the EU state aid intervention very seriously at first, simply arguing that the scheme did not fall under the EU definition of 'state aid' (MPE 2003). This was rejected by the ESA. Thereafter the MPE began arguing that the renewable support scheme had been designed with a view to minimizing societal costs, and thus no party would receive more state aid than needed to realize the production of renewable energy. However, this was not the main concern of the ESA, which instead probed into why Norway did not follow the extra-cost methodology prescribed by the guidelines. This methodology allowed the state to cover the additional costs of new renewable electricity as compared with conventional power production.

Two years followed in which the MPE tried to persuade the ESA to accept the cost-minimizing approach and disregard the extra-cost model (ESA 2005a; 2005b; MPE 2004). The ministry argued that hydropower was the conventional energy source in Norway, and because hydropower requires huge investments, such reference would allow for unreasonably high levels of support (MPE 2003:7; 2004).

Eventually, the ministry persuaded the ESA that the logic of minimizing societal costs made more sense than the extra-cost model (ESA 2005a). Then, after the ESA decided to base its ruling on this approach, they found that Norway did not in fact comply fully with this approach. The ministry was taken aback when the ESA issued a formal investigation of the Norwegian support practice on these grounds (ESA 2005b). As one interviewee noted: 'We should have gone into detail at first. Because we did not do that, we actually ended up making things harder for ourselves. Now we actually had to follow up what we had said. In fact, it did not function as well as we had said.' The MPE responded by ensuring that the actual practice of Enova became streamlined to the logic of minimizing societal costs. Finally, the ESA adopted this modified Norwegian support scheme in 2006 (ESA 2006).

Thus we see that adjustment to the EU regulations had the counter-intuitive effect of making Norwegian practice more estranged from the technology development logic of the state aid guidelines than before. The Norwegian approach was also very different from the feed-in schemes that diffused among EU member states at the time. Note that the biomass exception in the regulations on state aid shielded the renewable heating scheme from EU interference. Thus Enova had far more leeway for introducing adjustments and incremental changes in this scheme than in relation to electricity, so we do not see the same streamlining process in relation to district heating as we did with renewable electricity in the period 2000–2005.

All in all, developments in the price of electricity, and not support schemes for renewable energy, attracted the most political competition in this period. Precipitation was low in the winter of 2002/03 and because the Norwegian energy system was heavily dependent on large hydropower, the price of electricity soared. The consumers had to pay, while the power producers raked in large profits. Because the state and the municipalities owned most energy producers, this led to increased public revenues. The media, however, were brimming with consumer protests. Labour was among the parties that most forcefully blamed the conservative coalition government for the high electricity prices (Innst. S. nr. 167 2002–2003). Whereas stationary energy producers had used the first years after liberalization to make their hydropower operations more efficient, this peak in prices created new interest in the construction of power plants. Rising gas prices made that source of energy appear less profitable, but large onshore wind power became more attractive, due not least to the favourable wind conditions in Norway (see MPE 2008a).

While politicians were preoccupied with discussions over high electricity prices and gas power, the MPE administration was busy aligning Norway's support schemes with EU regulations. Now that Labour was in opposition, it had no problems with demanding the introduction of a renewables scheme to be paid by the consumers, and protesting against high electricity prices. That was to change, however, after Labour again took office in 2005.

Political attention to renewable energy was rather low in the period between 2000 and 2005; politicians were not strongly committed to any particular support scheme design, and tended to support the green certificate idea in order to communicate their symbolic support for renewable energy. In practice, the ESA process had far more impact on the development of supply to renewable electricity than the political deliberations on green certificates in this period. True, the Storting repeatedly supported green certificates, and negotiations with Sweden were initiated, but this did not lead to any actual changes in the support scheme. No high-level processes played out in relation to district heating in Norway. Rather, the support scheme resulted in a few district heating projects, and the criteria for funding changed somewhat as Enova gained practical experience. In the shadow of political competition centred on gas power and the electricity price, fiscal incentives regimes were developed for renewable heating as well as electricity during this early period, although the latter became more streamlined to the cost-minimizing logic than the former.

Phase 2, 2005–2010: Electricity Drama and District Heating Harmony

Renewable energy had still not become an issue of high political salience by 2005. True, negotiations on a common scheme were underway between Sweden and Norway, but political attention to the discussions was rather low. As we will see, the political status of renewable electricity changed radically during the coming years, while the renewable heating policy continued to develop without much political interference.

Let us first turn to the situation of renewable electricity in Norway. The 2005 elections brought a change of government, with the Labour Party forming a 'red/green' coalition with the Socialist Left and the Centre Party (former Agrarian). Interviewees report that renewable energy was not high on the agenda during governmental negotiations. However, the new government readily declared that it would continue to negotiate with Sweden on a common green certificate scheme (Soria Moria 2005). This came towards the end of a lengthy process with several years of assessments and close collaboration between Norwegian and Swedish civil servants. Interviews indicate that the creation of a unilateral Norwegian scheme was never seriously discussed, apparently because economic calculations showed that a common scheme would be far more cost-efficient.

Interviews show that it was only after the 2005 elections that Labour realized that a green certificate scheme might well mean increased electricity prices. Because green certificate schemes are financed directly by the consumers, the introduction of such a scheme will lead to an immediate increase in the price of electricity; the higher the renewable electricity target, the higher the cost burden on the consumers. Eventually, the scheme will serve to increase energy production, which in turn may eventually reduce market prices, but this effect is hard to predict (NVE 2004). Moreover, companies with the most profitable projects could reap far more profits from the scheme than what was strictly needed for it to become profitable. These features contrasted to the minimizing societal cost logic and would enable the opposition in the Storting to blame the government for possible future electricity peaks (even though the price of electricity was set by the market, not by the politicians). In an interview, one political appointee remarked that at this stage 'we were living in the shadow of the power price issue.' Another noted: 'Labour was scared to death by the protests against high electricity prices.'

Some politicians report they found it demanding to collaborate with the administration in the MPE on the certificate issue, because the ministry preferred the Enova cost-minimizing support scheme. One political appointee described the situation like this: 'They understood that they ought not to protest, but they did not always manage. They sat there moaning and groaning.' Others refer to the prime minister as having a central role in these discussions. According to one political player: 'Frankly, this was about the will of one person. It was all decided by the intervention of the prime minister's office.' Prime Minister Jens Stoltenberg is also described as not seeing 'any point in spending money on this when we have so much renewable energy to begin with.' Another prominent political actor underlined that the whole government gradually became more sceptical to joining with Sweden in a common green certificate scheme when they learned more about how it worked.

All along, the Swedish government had been clear in its demand for a high Norwegian target, on the grounds that if the coherent target became too low only the least costly projects would be realized—and these were to be found mainly in Norway, not in Sweden (Energimyndigheten 2005). Due to the arguments noted above, those sceptical to the certificate—politicians and administrative actors alike—wanted a low Norwegian target: the higher the target, the more likely was it that electricity prices would rise significantly and that commercial actors would reap big profits. Thus, the Norwegian government held out for a far lower ambition level than the Swedes were ready to accept—and negotiations between the two countries failed (MPE 2006a).

Neither the utilities nor the environmental organizations performed much lobbying during the most intense phases of negotiations between Norway and Sweden; interviews show that they were convinced that they had already won. Instead, they concentrated on competing for the best projects and enhancing their organizational capacities. Five of the ten largest utilities included wind power in their organizational structure, and many smaller companies created shared wind-power development subsidiaries (see Agder Energi 2007; BKK 2007; NTE 2004; Statkraft 2005). Shelving of the Norwegian–Swedish scheme came as a major blow to their aspirations. When they realized that their strategies had been founded on wishful thinking, they reacted with an outcry, and the environmental movement and MPs swiftly jumped on the bandwagon (see Innst. S. nr. 205 2006–2007:5).

The executive politicians were unprepared for these strong reactions. As one interviewee remarked: 'I was also a bit surprised over how much turmoil emerged. The industry had not given a clear enough message.' Because the opposition seized the opportunity to criticize the government, renewable energy suddenly rose to the top of the political agenda. As the political competition intensified, the red/green government initiated hasty efforts to sweeten the retreat from green certificate plans. It promised to develop an alternative scheme that would combine the concerns of all three parties in government: this scheme offered high operation support and was technology-neutral, like a certificate scheme, but would be financed by the state instead of the consumers (MPE 2006a). Interviews shows that it was only at this point, after having rejected the green certificate scheme, that political appointees in the Ministry of Petroleum and Energy realized that the EU guidelines on state aid severely constrained their ability to design an alternative scheme. The streamlining of the Enova scheme that resulted from the ESA processes made it impossible for the ministry to accommodate any of the new requirements without having to launch a new process in relation to the EU guidelines for state aid.

Interviews show that when the politicians turned to the ministerial administration to develop such a scheme they were met with the argument that the ESA would require a strict cost-minimizing logic. After months of internal deliberations, the government issued a draft of a hybrid scheme for the ESA that was primarily aligned to the logic of minimizing social costs (St. meld. nr. 11 2006–2007). However, rather than straightforwardly following the same criteria they had adopted in the former notification process, the ESA now referred to new precedents that had been created by Commission notifications of EU support schemes in the past two years (ESA 2007c). Thus, the Norwegian government realized that deliberations with the ESA could be lengthy and that it would take substantial time to develop a new scheme. This put the ruling coalition in a very difficult position. Interviews show, with green certificate schemes exempted from the ban on state aid, the politicians now decided to pick up the green certificate idea again. Note that this happened despite the low popularity of green certificate schemes in Europe. By now, 7 of the 27 EU member states (among them Sweden and the UK) had green certificate schemes, whereas 18 (including Germany and Spain) had various kinds of feed-in schemes (Commission 2008a).

According to one interviewee: 'It became evident that we had to do something. And then it was decided at a high level that it was a priority to engage in new green certificate discussions with Sweden.' The utilities had continued to lobby for a certificate scheme, but interviews show that they were near to giving up when the government suddenly changed its mind. As for the minority in the Storting, it had kept calling for such a scheme, and now lauded the government for 'finally coming to its senses' (see e.g. Innst. S. nr. 145 2007–2008).

At this stage, EU-internal discussions on renewable support had heated up again. In 2007, the Commission prepared to launch a draft for a revised renewables directive (Nilsson et al. 2008). Once more the Commission promoted a green certificate scheme; this time they had even designed a fully-fledged European scheme (Boasson and Wettestad 2013). This gave rise to even stronger protests, and once again the technology-development logic of the small renewable-energy producers prevailed (Commission 2008b). The new EU renewables directive permits member states to develop common certificate schemes, but they are in no way instructed to design market-reliant schemes (Directive 2009/28/EC: Art 11). While the first renewables directive transferred hardly any authority to the European level, the 2009 directive gives the Commission a stronger position. The new directive introduces a new binding target, requiring a 20 per cent renewables share in each country's energy system by 2020. This target encompasses all major energy-consuming sectors: electricity, heating and transport. Each member state is given an individual target, but because Norway is only a party to the EEA and not an EU member state, the Norwegian target was developed in separate negotiations at a much later stage. The European turmoil over the support-scheme-related content of the directive attracted scant attention in Norway. Only when the final directive had been settled did it start to get attention in Norway—but this happened too late to have much impact on the outcome of renewables policy in 2010.

New certificate negotiations with Sweden did not provide a quick fix. The scheme remained pending for three years, in part because of the new EU directive. Sweden is required to comply with the directive and to achieve its individual 49 per cent target, so the Swedish government wanted to check all details with the Commission. Moreover, lengthy new deliberations were needed on the details of the scheme. Not until late in 2009 did the governments of Norway and Sweden agree on the main principles for the new scheme, and in December 2010 they agreed on the main content (MPE 2010b, 2010c). The new scheme was put into operation in January 2012.

Turning to *district heating,* post-2005 developments were radically different from what we have seen in relation to electricity. Initially, the small-scale renewable actors promoted the inclusion of heating in the certificate scheme (see MPE 2005a). But as one interviewee noted: 'They did not really want a green certificates scheme, what they wanted was more state aid.' Thus, when the first green certificate negotiations failed, they swiftly shifted their strategy towards ensuring improvements of the existing scheme. The funds available for renewable energy and energy efficiency increased by 600 per cent from 2002 to 2009, with district heating the largest beneficiary by far (Riksrevisjonen 2010a:31). Particularly after 2006, applications for Enova support for district heating projects escalated, related to a new peak in electricity prices in 2005/2006. In contrast to the situation in 2002/2003, this price peak brought a boost in interest in district heating. Now 6 of the 10 largest utilities included heating in their governing structure (see Agder Energi 2007; Akershus Energi 2008; Eidsiva 2007; Hafslund 2007; Statkraft 2007). One interviewee noted: 'It worked as an eye-opener when Hafslund spent 260 million Euro to get 66 per cent of the shares in Viken district heating.' In addition, ambitious actors with no history as stationary energy providers appeared on the scene. These covered the whole value chain and launched bold growth objectives (see Solør Bioenergi 2009). Several interviewees remarked that the emergence of new actors led the traditional players to see the commercial potential in heating.

Between 2005 and 2010, the district heating programmes of Enova were re-organized several times (Riksrevisjonen 2010a:39). It seems that the various industry actors and Enova collaborated closely on the gradual improvements of the scheme. Interviews show that, as a result of the growing interest in heating, Enova launched dialogue processes with both the large utilities and the new smaller actors on how to improve the scheme. As district heating consists of many functionally disconnected systems, like several separate monopolies, the energy price is not readily evident (Havskjold and Halseth 2007). Because it is impossible to know exactly what the local price of heating will be in advance of project realization, Enova can follow a strict minimizing societal cost approach. Thus, in the process of improving its support schemes for renewable heating, Enova blended economic and technological criteria. For instance, the scheme was divided into more sub-groups. In that way, various different kinds of district heating projects do not compete with each other: infrastructure development and district heating construction compete only with similar kinds of projects.

Moreover, local heating is supported on the basis on technological criteria— so there has been a modest development away from economic criteria and towards technology criteria. This has not given rise to controversy. Most interviewees expressed a high degree of satisfaction with the functioning of the scheme. With exception of the rather steep increase in funding, there were no dramatic shifts in the support scheme for renewable heating, which developed incrementally throughout the decade.

By 2010, renewable energy had been established as a main priority among politicians and commercial actors alike. During much of the preceding decade, political discussions relating to renewable electricity had reflected the two opposing views of appropriate support to renewable energy: one aligned to the societal cost-minimizing logic, and the other reflecting the market logic. Somewhat surprisingly, there seems to have been little overlap between the two sub-processes. Towards the end of the decade, the political turmoil concerning electricity had calmed down, and it seemed as if both the fiscal incentive scheme for heating and the market measure for renewable electricity had gained wide acceptance.

The years from 2005 to 2010 provide considerable drama with respect to renewable energy, with a radical shift in policy. By contrast, district heating policy developed incrementally, with a steep rise in actual project funding.

5.4 ASSESSMENT

Norwegian politicians initiated policy development in relation to renewable heating as well as electricity, but the political field had limited impact on the actual design of the two strings of measures. Tables 5.1 and 5.2 present the mechanisms in operation in the two cases.

Table 5.1   Mechanisms in the case of renewable electricity

Phase ⇨ Social system ⇩	Phase 1, 2000–2005	Phase 2, 2005–2010
Political field	*Ministerial Governing*	*Politicizing*
Organizational field	*Turf Battle* Spider Importer Shrewd Lawyer	*Turf Battle* Spider
European environment	*Bottom up harmonization* *Unpredictable EU* *governing*	*Bottom up harmonization* *Unpredictable EU* *governing*

Table 5.2   Mechanisms in the case of renewable heating

Phase ⇨ Social system ⇩	Phase 1, 2000–2005	Phase 2, 2005–2010
Political field	*Ministerial Governing*	*Ministerial Governing*
Organizational field	Turf Battle/ Collaboration	Turf Battle/ Collaboration
European environment	*1000 Flowers*	*1000 Flowers*

The cost-minimizing, state-funding character of district heating measures mirrored the institutional logic of the MPE, whereas the market measure relating to renewable electricity reflected the logic of the utilities. Incremental change unfolded in relation to heating, but this took place within the original framework: the minimizing societal cost hallmark remained, even though some technology criteria were added. With renewable electricity, there was a radical shift from minimizing societal cost to a market measure. It seems as if the lack of entrepreneurial activity relating to heating provided rapid stabilization of this policy area, whereas the high degree of entrepreneurship in relation to electricity eventually led to policy displacement. The case of renewable heating is remarkably stable as regards the social mechanisms at work: no notable change in the mechanisms within any of the social spheres. During the decade the policy changed incrementally, driven by field-level developments, but the turf battle conflict was not effectuated. Note that the European environment situation differs in the two cases: stronger European mechanisms are at work in the electricity than the heating cases.

The Political Field: Politicians Unable to Escape Their Powers

Initially, the political field was in a Ministerial Governing mode: politicians had a certain normative commitment to renewable energy, but their engagement was characterized by 'garbage can' dynamics. Formal authority over renewable energy issues lay within the MPE. True, the Parliament's Standing Committee on Energy and the Environment repeatedly called for green certificate assessment, but the processes relating to supervising the Enova scheme, assessing green certificates and negotiating with Sweden were steered by the ministry. No other ministry had formal powers in this issue-area—unlike the case with respect to CCS, where also the Ministry of the Environment had formal powers.

The 2000 parliamentary decision was instrumental and spurred policy emergence with respect to both energy sources. Even though the decision was made in the shadow of the gas-power difficulties, it was fairly clear and concrete, involving two quantitative targets and the creation of the public enterprise Enova. The lack of political salience was no obstacle.

It seems as if the politicians simply acted on the impulses from the organizational field, paying scant heed to the actual difference between the two policy designs. The conflict between the logic of minimizing societal costs and the market logic simmered for years at organizational field level until it eventually created a major hurdle in the political field, resulting in politicizing of the issue in the second period. The organizational field-level actors had succeeded in politicizing the issue, thereby creating the need for a political solution to the conflict. Because the political commitment to the issue had been a 'garbage can' situation for years, neither the politicians in government nor the opposition had clear and entrenched positions on the issue at the start of the second period in focus here.

Up to this point, the actions of politicians had been characterized mainly by strategic thinking in relation to political competition on other issues, not renewable electricity as such. For instance, electricity prices had greater normative importance than renewable electricity. Due to the liberalization of the energy sector, politicians now had few means of controlling the energy price, but they still got blamed if they did anything that could affect the price. Because this was an issue of high public saliency and media attention, the politicians were nervous: they did not want to risk losing political credit. Calls for a green certificate scheme merely served as a rhetorical ploy that allowed the opposition—no matter which parties happened to be in opposition—to complain about the efforts of the government. Eventually, however, these symbolic actions yielded highly tangible consequences: the failure of the initial attempt to develop a certificate scheme gave rise to a major political controversy that pressed the politicians into finding a viable solution. In 2006, for the first time, the politicians found themselves faced with a situation where they had to make a binding decision on the certificate scheme.

We may interpret the refusal to accept the high ambitions demanded by Sweden as a result of the Norwegian government's misreading of the situation; not grasping the magnitude of the uproar this would lead to, they failed to take it into account. However, we should also bear in mind that the opposition might have reacted very differently if the government had accepted the agreement with Sweden. In that case, the conservative parties in the Storting could just as well have attacked the agreement and blamed future price peaks on the government. Nonetheless, we can conclude that because the government initially paid little attention to renewable energy, they ended up in this situation. Had it been in their power to do so, they would have developed a quick fix (such as a hybrid scheme) as soon as they realized this failure. The conflict between the demands from the industry and the ministry, and the importance of EU regulations on state aid in this respect, served to obstruct such solutions.

A vicious circle of conflict created increasing tension and in the end a radical policy shift on renewable electricity. Note that the politicians did not develop firm preferences until many years after they had made the formal political decisions that promoted green certificates; in line with the original garbage can theory, their cognitive beliefs came as a consequence of their actions, not the other way round (see Cohen et al. 1972).

Renewable heating differs from renewable electricity in the sense that political competition never emerged in relation to this issue-area: because no one posed strong objections to the Enova scheme on the organizational field level, the issue never became politicized. Politicians merely endorsed the incremental changes that occurred at field level, also the introduction of technological criteria and escalation of project support after 2005. The lack of political tension in relation to heating seems to have created good conditions for deliberations between different field-level actors and incremental change in scheme characteristics. The incremental change and eventual success with respect to industry change and project realization gave the politicians no incentive to demand changes in district heating.

There was a striking lack of political entrepreneurial activity. Not even when renewable electricity became politicized did the politicians launch strategic activities directed at overcoming structural and institutional barriers. One key reason is that political salience plummeted as soon as the final decision to join the Swedish scheme was made.

On this backdrop, we may conclude that the political field played a key role in initiating the two strings of renewable measures, but had limited impact on the actual characteristics of the measures. In both cases, politicians responded to developments within the organizational field. With renewable electricity, the political field had to deal with the two conflicting logics (market versus minimizing societal costs) at organizational field level, but this conflict was not mobilized in relation to district heating. Note that lack of firm political steering underpinned policy instability with respect to renewable electricity, while it underpinned stability with respect to heating.

The difference in effect can be understood only when conditions in the organizational field are taken into account.

The Organizational Field: Turf Battle Mechanism Effectuated for Renewable Electricity

The organizational field of stationary energy production in Norway consists of a relatively small group of commercial actors and various state agencies under the MPE. Within each of these two groupings, one organization clearly has the upper hand: the ministry on the state side, and Statkraft among the commercial actors. Both have considerable structural resources at their disposal. The MPE has authority to take key political decisions and has highly competent employees; Statkraft has authority to take investment decisions in relation to its own projects, and also contributes to shaping how other commercial actors plan their investments, particularly by influencing the work of the energy business association. Moreover, Statkraft has superior access to information concerning market conditions and technological developments. However, the two sides also have considerable contact, and Statkraft is fully owned by the Norwegian state—so the structural resources of the field are in fact rather centralized.

The character of the field diverges from the descriptions provided by segmentation scholars and early neo-institutional sociologists: it is not controlled by any one actor, nor is it totally dominated by any one institutional logic. It does not fit the projections of the neo-pluralists either: it is not characterized by free competition between a wide group of actors with differing views and interests but similar structural power basis: it is dominated by two equally powerful organizations. When it comes to professional logics, energy market liberalization pushed Statkraft and the ministry apart, because the logic of societal cost minimization now came to dominate within the ministry, and market logic within Statkraft. Because of these differences in logics, conflicts between the two are not infrequent.

The organizational field seems to have played a minor role relating to emergence of the two renewables policies in focus here: electricity and district heating. Already in 2000, Statkraft and other utilities had begun to develop an interest in wind power. This may have sustained the political initiatives that were taken at the time, but the power producers do not seem to have been the driving force. Neither of the two dominant field actors showed any enthusiasm for district heating at this point. Hence, both of the two policies emerged despite low engagement at field level. Initially both policies on renewables were aligned to the minimizing societal cost logic, in line with the dominant logic of the ministry.

Not long after the decision to develop a cost-efficient support scheme for renewable electricity was taken, the diverging worldviews of the ministry and Statkraft clashed. As the industry's interest in wind power grew stronger, the conflict became more pronounced. Field-level actors were not able to cope with this conflict themselves, so Statkraft soon prompted politicization of the issue. Thus the commercial actors found themselves dependent upon a political solution. The lines of conflict in the political debate largely mirrored the arguments of the two main actors within the organizational field.

The conflicts endured and the commercial actors refused to align to the actual support scheme; they declined to apply for project funding and instead decided to wait for the shift towards green certificates. These tensions over the design of the electricity support scheme created an enduring uncertainty that obstructed actual investments and hindered governmental and commercial actors in concentrating on all the technical details involved in getting this particular policy measure to function in practice.

Because the field was dominated by the same social mechanisms in both cases, it cannot fully account for why the commercial actors were willing to accept a societal cost minimizing scheme for renewable heating, but not for renewable electricity. It was entrepreneurial efforts, not the turf battle mechanisms as such, that created a deadlock in the policy developments regarding renewable electricity. Entrepreneurship was exercised in two different settings: first by Statkraft and other green certificate promoters, who put the green certificate issue on the agenda in the first place; second, by the MPE, which re-directed the effect of the EU regulations on state aid.

In the early years of the new millennium, Statkraft, supported by environmental groups, exerted Spider entrepreneurship, encouraging the opposition to instruct the government to assess the development of a green certificate scheme. The politically skilled employees of Statkraft circumvented the ministry and approached the Storting directly; they were able to exploit the thirst of the political opposition for symbolic and ritual criticism of the government. However, such entrepreneurial activities were not set in motion with respect to district heating, only green certificates.

Green certificate promoters also exerted normative entrepreneurship of the Importer kind. Already in 2000, Statkraft argued that green certificates were the preferred European choice: Norway ought to learn from Europe and jump onto the bandwagon. This line of argument had considerable strength and contributed to successive endorsements by the Storting.

However, the promoters of green certificates became blinded by their own swift success. Believing that they had already won, they failed to explain to the politicians the nitty-gritty details of what a green certificate scheme would actually imply. And so it was that green certificates remained a mere symbol of a good renewables scheme in the eyes of the politicians. The entrepreneurs, for their part, did not recognize the lack of opposition and apparent interest in details as a sign of limited political interest in renewable energy in general. To be sure, their efforts led to the institutionalization of green certificates within the organizational field of stationary energy production, but this process came to a halt in the political field. It is particularly noteworthy that the entrepreneurs failed to realize that the other dominant actor in the organizational field—the ministry—was actively countering their efforts. All the same, despite this lack of initial success, Spider and Importer entrepreneurship were preconditions for the later shift to green certificates.

The structural as well as the institutional entrepreneurship of Statkraft and companions drastically decreased in the period 2004–2006, which coincided with critical junctures in the initial negotiations with Sweden. This resulted from a strategic miscalculation: the entrepreneurs misread the situation and acted upon wishful thinking. This indicates that Statkraft was used to having things its way and did not fully understand that the green certificates idea had not manifested itself politically. After it became evident that the negotiations had failed, the whole set of utilities and environmental organizations set about engaging in entrepreneurial efforts. Yet, it seems that it was primarily the problems relating to getting ESA acceptance for a hybrid scheme that led the Norwegian government back onto the green certificate track, not this second wave of entrepreneurial activity.

A different kind of structural entrepreneurship was exercised by the administrative staff in the Ministry of Petroleum and Energy, related to the EU guidelines on state aid. These guidelines are characterized by Unpredictable EU Governing: even though it is highly formal and binding, this EU legislation contains considerable room for interpretation. After several years, the MPE succeeded in persuading the ESA that the Norwegian energy system was so special that the regular EU procedures for state aid could not be applied, and the ESA ended up supporting the minimizing societal cost approach. Political executives in the ministry had the authority to intervene in this process, but they failed to pay much attention. This appears to have resulted not from any deliberate effort of the administration to keep matters secret: it was more that the political appointees in the ministry already had their hands full and lacked time and energy to devote to the ESA process. Nor is it surprising or strange that the political leadership should choose not to engage in an issue that had been presented as a technical-juridical hurdle of only minor political importance.

The entrepreneurship performed by the administrative staff in the ministry fuelled tensions even higher with respect to renewable electricity—a fact which further contributes to explaining the differences between the two sub-cases. The extra-cost model of the EU state aid guidelines was far more flexible than the approach championed by the ministry, which relied on rather rigid calculation procedures. Thus the entrepreneurial success of the ministry created a new formal constraint, one that later undermined the politicians' ability to create a hybrid scheme for renewable electricity.

Because the green certificates measure was exempted from the EU rules on state aid, it entailed fewer difficulties for national policymakers than did other measures. Eventually, the problems related to the regulations on state aid helped to promote the shift to green certificates, but there is reason to believe that the transition would have been smoother and have come earlier if the certificate entrepreneurs had played the state aid card more actively. The Shrewd Lawyer entrepreneurs were indeed able to alter a formal EU steering signal, but the negotiation success of the ministry eventually had a highly paradoxical effect: it actually helped to strengthen the authority of the EU over Norway's renewable electricity policy.

We may conclude that the entrepreneurial promotion of green certificates was successful in the sense that this solution was endorsed in several formal political decisions, but due to half-hearted engagement it took 10 years until the entrepreneurs could reap the profits of their efforts. Because the two groups of structural entrepreneurs promoted different policy solutions, the joint effect of their efforts was the emergence of a major conflict, which led to instability and eventually a shift towards a market measure.

What then of district heating—why did the potential institutional conflict never materialize there? Firstly, the power producers seem to have been were hesitant towards to this energy carrier because heating was very different from the commodities the energy producers were accustomed to dealing with. Due to the initial lack of interest in the subject, Statkraft never voiced objections to the Enova scheme. (Actually, Statkraft and other large utilities became aware of the potential profitability of district heating only after the scheme had spurred significant field-level activity.) Second, the special market situation for heating can help to explain why Enova was able to design a scheme that would provide significant profits for the market actors: because the price of heating is created in local markets, the energy price on which support calculations are based is always negotiable. By contrast, the electricity price is given by the energy stock exchange. Because the government simply cannot control the profits as they can in relation to electricity, it is not technically possible to design a streamlined societal cost minimizing scheme for heating. This ambiguity provided the field-level actors with leeway to negotiate the functioning of the scheme—but these deliberations were carried out among technical experts, not by the top management in governmental organizations and the utilities. That led to the introduction of technical criteria which in turn brought a greater number of projects, because also more costly and small-scale heating projects could now be granted funding. All the same, the societal-cost character remained strong.

Third, the slow but steady increase in the number of funded projects served to spur commercial interest in the scheme. No entrepreneurial activities obstructed the steady path-dependent development, and the lack of conflicts served to ensure stability. The Turf Battle mechanism was never effectuated with respect to renewable heating; rather, actual investments and gradual adjustments of the scheme created positive feed-back effects, continually reinforcing the policy along the initial path.

Thus, we see that the organizational field is less relevant for explaining the emergence of these policies, but is central in explaining the actual character and stability of the policy measures. Field-level social mechanisms cannot fully explain why the potential field-level conflict materialized in relation to one energy source only: it was entrepreneurship that aroused conflict in relation to electricity and not heating. As we will see, European developments contributed to make it particularly challenging for the political actors to create stability in relation to renewable electricity, whereas district heating was not affected.

European Environments: Creating Counter-Intuitive Effects

Several European countries were ahead of Norway, having already developed thriving policies for renewable electricity by the year 2000, but these national practices attracted little attention at the time when Norway was developing its policies on renewables. The European Environment had a bottom-up harmonization character, with feed-in schemes diffusing rapidly in Europe. Feed-in schemes and the logic of technology development gained increasing strength in the wider European environment, in EU renewables policy and state-aid regulations, but it was the market initiatives from the Commission and the large European utilities that attracted attention in Norway. The green certificate idea gained strength there, while it became less popular in Europe. The fierce EU conflicts between feed-in supporters and promoters of green certificate schemes (with the eventual defeat of the latter) might have been expected to feed into the development of Norwegian policy, but this hardly happened. That clashes with dominant accounts of policy diffusion among neo-institutionalist scholars, who hold that a policy idea will diffuse more efficiently, the more actors who adopt it (see e.g. Finnemore and Sikkink 1998; Sahlin and Wedlin 2008).

EU regulations on state aid also played into the renewable electricity debate, whereas renewable heating was shielded because biomass was exempted from the ban on state aid. The state-aid guidelines give the EU considerable authority over the development of national policy on renewable energy. Even though the environmental state-aid guidelines follow the logic of technology development, alignment to these rules resulted in streamlining the cost-efficiency character of Norway's support scheme for renewable electricity. Hence, the European environment has had paradoxical consequence also in this respect. It is evident that entrepreneurial activities were important in the transformation of the original EU steering signals also in this respect.

The development of district heating in the other Nordic countries may have inspired Norwegian politicians before they made their decisions in 2000, but this does not appear to have been decisive to the outcome. Moreover, renewable heating was only a Nordic, not an all-European, phenomenon at the time. Heating attracted broader EU attention only some five years later, but this does not seem to have affected the Norwegian support scheme in any way.

5.5 CONCLUSIONS

This chapter has examined the development of two Norwegian policies of renewable energy: for electricity and for heating. We have seen how both emerged as cost-minimizing state aid (fiscal incentives), but the electricity support converged to a market measure. The multi-field assessment has shown that the political field initiated policy developments, but that the conflict in the political field mirrored organizational field conflicts. All in all, the organizational field dominated the development of policy, and the political field had limited independence. However, it was differences in entrepreneurial activity from organizational field actors that ensured that the field-level conflict between minimizing societal costs and market logic became politicized in relation to electricity but not heating; differences in entrepreneurial activity explain the variation in policy outcome and change between the two issue-areas.

The European environment provided far more opportunities for entrepreneurship with respect to electricity than district heating, and Norwegian entrepreneurs exploited this in a way that fuelled national conflict and eventually led to a policy shift. The institutional conflicts that emerged in relation to renewable electricity resulted from two different lines of entrepreneurship: Statkraft, which launched a campaign for a green certificate scheme; and the Ministry of Petroleum and Energy, which acted as an entrepreneur because EU regulations on state aid required it to take action. This conflict served to postpone the establishment of new renewable electricity activities, but eventually ensured a shift in the character of policy. The success of the green certificate entrepreneurs came at a very high cost: investments in renewable electricity were severely delayed until political clarification was given.

Renewable heating had low salience. Thus, entrepreneurship did not emerge, and potential institutional conflicts were not effectuated. This enabled actual implementation of the policy measure and incremental changes that supported the economic interests of the industry. Here we see a path-dependent development that was to result in an overly cost-minimizing policy outcome in 2010 as well as a high level of stability and high degree of actual industry change. This indicates that high political saliency is not a precondition for the emergence and development of effective climate policies.

Our assessment of Norwegian renewable energy policy shows that entrepreneurial strategies that result in politicizing can prove risky: the politicization of electricity was a de-stabilizing factor, particularly because it took so long from when the entrepreneurs started to induce the conflict until it gained a competitive character and the politicians actually devoted energy and time to solve it. It may be asked whether the entrepreneurs would have been able to act in a way that led to a smoother renewable electricity process if they had been more aware of the social landscape in which it was embedded. The fact that actors lacked firm preferences with regard to policy design was a blessing for the development of district heating policy, while the relatively high level of entrepreneurship as regards renewable electricity hindered the development of a stable policy.

NOTES

1. The Norwegian Water Resources and Energy Directorate (NVE).
2. For example, the former CEO of this association is now employed by Statkraft, while the new head of the association came from a position in Statkraft.

6 The Strength of a Pluralist Organizational Field
Energy Policy for Buildings

6.1 INTRODUCTION

Housing was a central issue in European welfare-state policies in the early decades after the Second World War (Kemeny 2001). Later, de-regulation of national housing markets served to reduce political control over building construction. Since 2000, buildings have attracted new attention in terms of energy performance, in Norway as in most countries throughout Europe. Despite the brief political history, the range of policy measures targeting buildings has become broader than in any other area of Norwegian climate policy.

The concept of 'high energy-performance buildings' is still in the making. Definitions abound, and the new concept also goes under many names, such as 'passive houses', 'low-energy buildings' or 'zero-energy buildings'. Buildings are responsible for more than one third of all Norwegian energy consumption (Thyolt 2006). High-performance buildings have a building envelope with good thermal qualities (thick insulation, scant energy leakage), the internal distribution of energy is highly efficient, and low-scale renewable energy-producing equipment is installed on site (Dokka and Hermstad 2006). Through proper design, energy equipment and energy management, buildings can be transformed from energy consumers to energy producers. But here new skills are needed, and close collaboration between all involved actors, from building designer to electrician, is essential.

In the course of only a decade, an impressive set of energy-policy measures directed at buildings emerged in Norway—and without the public controversies that often unfold in relation to the emergence of new policy areas. This chapter asks: *Why did the Norwegian energy policy for buildings emerge? And why is it so diverse, consisting of four different policy measures?*

The organizational field of building construction comprises various actors with highly different worldviews, from architects to plumbers and roofers. Moreover, many governmental organizations relate to construction activity, none with overarching responsibility. Thus the organizational field is characterized by scattered structural powers and many institutional logics.

Political science segmentation approaches and the classic new-institution contributions all portray this as a situation in which the organization field will have little political clout. In contrast, pluralist and network approaches hold that, as long as field-level actors are good entrepreneurs, they will still be able to influence political decision-making. This chapter will argue that the diverse character of the policy reflects the pluralist organizational field; and hence the nature of the field has been very important for policy development—far more so than the political science literature would lead us to expect.

6.2 POLICY OUTCOME: DIVERSE AND INCONSISTENT

The energy policy for buildings covers a whole range of measures: technology standards, governmental industry development, market instruments and fiscal incentives. All possible combinations of indirect and direct state steering and specific technological development and diffusion of low-cost technologies are present in this policy outcome—it exemplifies an 'All in One' climate policy.

First, Norway's State Housing Bank developed *governmental industry-development measures*. This state-owned financial institution works together with dwelling producers on developing low-energy buildings, granting state aid to innovative energy-enhancing techniques and technologies for building construction. State aid and information campaigns are central measures (see e.g. Lavenergiboliger 2009). The Housing Bank grants support on the basis of technological, not economic criteria (Husbanken 2003). In 2006/2007, close to 40 per cent of its loans were for dwellings with improved energy qualities (Husbanken 2006). In 2003/2004 it developed the objective that all new buildings in 2010 should require half as much energy as that prescribed by the 1997 building code (KRD 2005:18, St. meld. nr. 23 2003–2004:19). This objective has not been altered since then.

Second, the Ministry of Local Government and Regional Development has developed a *building code with technology standard characteristics*. The energy requirements of this code regulate which techniques and technologies may be applied in building construction. New energy requirements were introduced in 2007, aimed at ensuring that new and renovated buildings used 25 per cent less energy than required by the 1997 code (KRD 2007). Whereas the previous building code had merely regulated the thermal quality of some construction components (floor, walls, roof), the new code regulates the totality of features that contribute to a building's total energy consumption; it is founded on a holistic understanding of the joint energy performance of all the characteristics of the building. A top priority is high thermal quality of the building shell, entailing a high-density construction where all building components (roof, floors, walls, windows etc.) must meet certain standards as to insulation. The building code does not allow poor thermal quality of the building shell to be compensated by installing in-house energy-producing equipment, such as heat pumps and solar panels.

The regulations are to be made more stringent every five years, aiming towards passive house standard (or some other demanding holistic standard) in all new buildings from 2020 (Innst.S. nr. 145 2007–2008:25). In addition, 'the building design shall ensure that a substantial part of the heating can be covered by other energy supply than electricity and/or fossil fuels' (KRD 2007). Oil-burning furnaces are prohibited in new buildings, and minimum 60 per cent of the heating in buildings larger than 500 m² is to be based on other energy carriers than electricity or fossil fuel (KRD 2010). For buildings of less than 500 m², 40 per cent is sufficient. This requirement may be set aside if calculations show that it will lead to extra costs when the full lifetime of the building is taken into account.

Third, *the Enova state aid scheme belongs under fiscal incentives*. Enova is a state enterprise that primarily offers state aid to piecemeal low-cost energy-efficiency improvements. Thus Enova does not apply the holistic approach found in the other measures, but focuses instead on specific building components, mainly targeting the largest commercial actors in non-residential construction (see Enova 2006, 2009; Riksrevisjonen 2010a:44). In 2009, Enova had as an objective 18 TwH new energy from 2000 to 2011 (Riksrevisjonen 2010a:27), to be met by promoting improved energy efficiency or the development of renewable energy. In 2008, Enova spent 20.7 million Euro (19 per cent of its budget) on energy purposes related to construction and building (Riksrevisjonen 2010a:32). Enova offered investment support to comprehensive energy agreements with major building developers/managers (implying the actors committed to achieve a certain volume of saved energy) and energy-efficiency investments in buildings (Enova 2009). Societal cost minimizing criteria determine the kind of support that is given, not technological criteria. Only those applicants that produce the highest energy surplus seen in relation to the support needed will be granted state aid (Riksrevisjonen 2010a:58–59).[1] No projects that are profitable at the outset will be supported.

Fourth, *the energy certification scheme for buildings* operated by the Energy Directorate is a *market instrument* (NVE 2010b; Ot. prp. nr. 24 2008–2009). All large non-residential buildings and other buildings that are rented or sold must be certified. The certificate will disclose information to the building market that may affect pricing signals. The energy certificate consists of two scales: one for the energy quality of the building, and one for heating solutions in the building. Calculation of the former differs from the holistic calculation approaches of the building code by favouring high-quality energy (electricity) over the low-quality energy provided by district heating or on-site bioenergy burners. Moreover, and in contrast to the building code, the energy certificate scheme allows building owners to compensate for a poor building shell by installing heat pumps and other on-site energy production. To some extent, the heating scale compensates for this difference. Energy experts are to be involved in the certification of non-residential buildings, while residential building certification is to be ensured by the building owner, facilitated by a web tool.

None of the four measures has aroused much controversy. The number of policy measures has remained rather stable since 2005, although it took many years from when the government promised to develop a certificate scheme for buildings until it was actually launched. The magnitude of measures is impressive, but the inconsistency and variation is puzzling. The next section presents the empirical development of the 'all-in-one' energy policy for buildings in the 2000s.

6.3 FROM NON-ISSUE TO CLIMATE HEAVEN

Phase 1, 2000–2005: Establishing Buildings as a Climate-Policy Issue

Because most buildings are heated by renewable hydroelectricity, they represent a lower share of the national carbon emissions in Norway than elsewhere. All the same, the increased saliency of climate issues has led to a greater focus on the energy performance of buildings. After all, if buildings are more energy-efficient, that should release electricity that can be used to replace fossil fuel used elsewhere. Norway had hardly any energy policy for buildings in the 1990s (Farsund 2000; Hubak 1998). It was not until the period 2000–2005 that the energy performance of buildings emerged as central to national energy-efficiency policy.

The first Norwegian energy-efficiency objective was introduced in March 2000, the same day as the government resigned over the gas-power issue (see chapter 4). On that day the Storting decided that it was an objective 'to reduce growth in energy demand more than business as usual' (St. meld. nr. 29 1998–1999, my translation). Further, it was decided to create a state enterprise, Enova, to administer an aid scheme based on cost-efficient support to specific energy technologies. Even though this represented a significant political shift it attracted little attention—the political debate was dominated by the gas-power struggle. The parliamentary decision in 2000 did not pay much attention to the energy performance of buildings, apart from a reference to the routine revision of the energy requirements in the building code. Due to this ambiguity it was not obvious that the decisions of 2000 would have any implications for Norwegian building construction policies at all. Moreover, it was highly unclear which ministry was to be responsible for taking the matter further: political responsibility for building construction is shared among the Ministry of Local Government and Regional Planning (responsible for housing policy and the building code), the Ministry of the Environment (responsible for physical planning), the Ministry of Trade and Industry (in charge of construction-related industry policy) and the Ministry of Petroleum and Energy (MPE) (for all energy-related issues) (KRD 2009).

Four different and loosely coupled processes relating to the energy policy of buildings unfolded during the next five years. Here we will look at the processes relating to the Housing Bank measures, energy requirements in the building code, the Enova scheme, and the energy certification of buildings, in turn.

We begin with *the Housing Bank* processes. Even though the parliamentary decision made no mention of the lending activity of the Housing Bank, the Housing Bank promptly responded to the new overarching energy-efficiency target by starting to promote low-energy dwellings (see KRD 2005). It developed its own energy portfolio through close collaboration with technological research communities, and also participated in international research projects. In addition it worked closely with Norwegian dwelling producers. These methods of operation must be understood in light of the Housing Bank's central position within the organizational field of building construction in Norway. This governmental institution had been created with the objective of solving the housing shortage after the Second World War (Guttu 2003; Pedersen and Steen 2005, Reiersen and Thue 1996). At this point, the technology development logic dominated the field, championed by architects and engineers—which for decades were the leading professions within governmental as well as private organizations (Pedersen and Steen 2005). In addition to building codes, technical standards developed by research institutes and the Housing Bank sought to ensure uniform standards for all dwellings; homes were to be affordable and solid, but not luxurious (Johnstad 1993:21, 162; Reiersen and Thue 1996).

In the 1980s, the government introduced the market logic into building construction. Interest rates were to be determined by market forces, and detailed technical requirements from the Housing Bank were scrapped. In parallel, financial investors acquired family-owned construction contractor firms, and economists replaced engineers in the top positions (Pedersen and Steen 2005). The biggest construction contractors were listed on the Norwegian stock exchange and were later acquired by Swedish corporations (NCC 2008; Peab 2008; Skanska 2008). Moreover, buildings had become popular investment objects. Real estate became traded goods, often with high ownership turnover. Insurance companies and banks turned to real estate, and eventually real-estate development became a specific business (Foreningen Næringseiendom 2009).

As the financial market logic gained in salience, attention shifted from standardized and universal buildings toward buildings constructed in line with individual preferences (Reiersen and Thue 1996; Torget 2004). Buildings, both dwellings and non-residential, increasingly served as visual expressions of identity. This implied fundamental normative change: whereas architects had taken a societal planner role before, they now came to see themselves as artists (Guttu 2003; Ryghaug 2003). Their chief aim was to create impressive modern buildings, not to fulfil specific functional objectives. This emphasis on the visual impression of buildings enabled architects to focus on market niches that offered higher profitability. The new architectural role fit hand in glove with the new market logic.

All the same, the market logic never became fully developed among construction actors. Most interviewees characterize the industry as conservative, and underline the prevalence of short-term perspectives and last-minute adjustments to governmental regulations. Few commercial actors in the field are public limited liability companies, so they do not have to relate to strict financial logic. With the exception of brief bonanzas, most actors strived to make enough profit to survive (Johnstad 1993; Pedersen and Steen 2005; Riksrevisjonen 2010a:44). Many interviewees underline that, due to small size, most companies lacked the capacity for strategically maximizing long-term profits. In addition, some elements of the old regime endured, like the technology standards in the building code and the Housing Bank itself. True, the Bank shifted its focus away from uniform standards for all building towards specific qualities, such as environmental qualities (Bachke 2003; Husbanken 1995:16; St. meld. nr. 23 2003–2004:59)—but its lending activity was still based on technical criteria, not economic calculations. It was the technical qualities of the building that determined whether it was eligible for support, not the costs entailed in production.

In the late 1990s, the Housing Bank became engaged in international research projects that promoted high energy-performance buildings. This brought the Bank into contact with a small group of European architects and technological researchers who advocated a holistic perception of the energy performance of buildings (Keulenaer and Gerwen 2008). The 'passive house' became a particularly famed practical application of this new approach (Halse 2005). With their solid insulation, passive solar heating and coherent energy planning, these buildings hardly need heating or cooling. Several thousand such dwellings were constructed in Germany and Austria, followed by other countries like Sweden (Halse 2005; Passivehuscentrum 2009). Interviews show that the new conceptualization of buildings was diffused primarily among experts, not among the many national industries. Moreover, the holistic approach served more as an inspiration than an actual model.

In contrast to the new holistic approach among experts, piecemeal approaches to energy-efficient buildings reigned in European building codes in the 1990s: only certain parts of the building, such as the thermal quality of the roof, walls and windows, were regulated (Visier et al. 2003:7). Technical requirements differed from country to country, but the underlying logic was the same: the government regulated the minimum technological standard for building construction, not the societal costs related to these construction efforts. Moreover, housing as such was at the centre of attention, and most measures were directed at improving the welfare of the people (Boelhouwer and van der Heijden 1993; Kemeny 2001). The holistic perspective on

energy performance represented a re-definition of the technology development logic: from welfare to energy issue, and from piecemeal regulation of
certain aspects of the building to a coherent technological approach. Yet
there remained considerable room for local interpretation. Many different
definitions of high energy–performance buildings emerged, giving differing
weight to on-site renewable energy generation, carbon emissions and thermal qualities (Jensen et al. 2009:12–13).

The Housing Bank and the technological research institutes were in the
vanguard of bringing the new European trend into Norway. Drawing on
their participation in international research projects, the Housing Bank and
the research institutions constructed several demonstration buildings (see
Støa et al. 2006; Thyolt 2006). This practical experience enabled the Bank
to present a Norwegian 'low-energy house' concept, adjusted to Norwegian
construction practice and with less strict energy requirements than the Ger-
man 'passive house' (Dokka and Hermstad 2006; Thyolt 2006:18). Moreover, the Norwegian experts focused more on the building envelope and
less on renewable energy than elsewhere in Europe (see KRD 2006b). Thus
the Norwegian conceptualization of high-energy performing buildings was
developed not merely through theoretical exercises but also by practical trial
and error. Moreover, the Housing Bank engaged in dialogue with the Ministry of Local Government and Regional Planning, the research communities,
the dwelling producers and to some extent also the Office of Building Technology and Administration. Like the Housing Bank, this agency (hereafter
'the Building Agency') is a subsidiary of the Ministry of Local Government
and Regional Planning. Interviews show that the ministry was positive to the
Housing Bank's initiatives, eventually also endorsing them in the ministry's
environmental action plan (KRD 2005).

Many of the same persons and organizations that participated in the
Housing Bank process were also engaged in the *revision of the energy
requirements in the building code*. The Building Agency was in charge of the
revision. The building code applies to all building construction endeavours,
but no governmental organization is directly involved in the construction
process in the way that the Housing Bank follows building construction.
While it is the government that develops the building code, it is up to the
research communities and the national standardization organization to
develop the more detailed requirements as to building standards. Local governments approve specific building projects and monitor compliance with
the building code.

Also the building code process had a certain EU policy aspect, in that
the European Energy Performance of Buildings Directive (EPBD) targeted
national building codes (Directive 2002/91/EC). By the turn of the millennium, the European Commission was repeatedly arguing that the building
stock was the largest energy-consumer and CO_2-emitter in Europe, and that
enhanced energy performance could reduce the EU's major energy deficit
(see Commission 2002, 2005c). In 2002, the EU adopted the EPBD. The key

term in that directive—'energy performance of buildings'—was a new invention that reflected the cognitive shift among experts. The EPBD promoted a holistic approach to the energy performance of buildings, requiring member states to take into account the thermal qualities of the building envelope, the efficiency of installed equipment (such as ventilation and tap water) and applied energy source. It was made mandatory for member states to include energy requirements in their building codes, targeting buildings over 1000 m² (Directive 2002/91/EC: Art. 3, 5). In focusing on large buildings, the directive differed from the traditional member-state focus on housing. No minimum thresholds were specified, but regulations were to be reviewed (implicitly implying strengthening) at regular intervals of maximum five years (Art.4).

In Norway, the Building Agency first asked the same researchers who had been involved in development of the Housing Bank measures to assess how energy requirements could be strengthened. These based their proposal on the same holistic understanding of the energy performance of buildings (see Thyholt and Dokka 2005). However, interviews show that the EU directive played only a minor role in the revision processes. Political appointees claimed not to have read the text of the EPBD, and the administrative staff argued that they paid scant attention to the directive during the course of the processes. For instance, they proposed energy requirements for all buildings, not only large buildings as proposed by the EU; and the methods for energy measurement were tailored on the basis of Norwegian practical experience, not descriptions in the directive. Interviews also show that Norway's Conservative/Liberal government in 2004 refused to launch new energy requirements prior to the 2005 elections. Thus the revision was put on ice. As we shall see, the new government that entered office in 2005 was to view this issue from another angle.

Let us now turn to the measures that were developed by the Ministry of Petroleum and Energy—first the Enova scheme and thereafter the energy certification of buildings. *The Enova scheme* that the Storting initiated in 2000 was up and running by 2001. Because buildings represent almost 40 per cent of the Norwegian energy consumption, Enova decided to develop a programme that targeted their energy performance. Initially it had an overarching aim of ensuring 10 TwH new or saved energy by 2010 (MPE 2002; Ot. prp. nr. 35 2000–2001). The target gradually became more ambitious, and available state aid funds increased incrementally during the decade (Riksrevisionen 2010a). Enova was established as a public enterprise. It was governed by contracts given by the ministry, but the government could not intervene in how Enova managed its daily operations. The energy-efficiency practices of Enova were not subject to any heated political debates or instruction.

The MPE supervised the activity of Enova. This ministry has general energy expertise, but not specifically on the energy performance of buildings. In line with the dominant principles in Norwegian energy policy in the early 2000s and the societal cost minimizing logic of the energy divisions in the ministry (see chapter 5 and Farsund 2000), the ministry instructed Enova to support only those energy-efficiency measures that were cost-efficient—meaning only the measures that needed the least support in order to realize the greatest energy gains. Moreover, Enova had a piecemeal approach to energy efficiency: it supported the improvement of the energy quality of the various parts of the building, without applying a holistic approach that took into account the energy performance of the whole building. Interviews show that the technical researchers so central to the measures developed by the Ministry of Local Government and Regional Development were hardly involved in the development of the Enova scheme.

The last measure, *energy certification of buildings,* was directly related to the EPBD. In addition to building code requirements, the directive included this market measure, which was to apply to new buildings as well as the existing building stock (Directive 2002/91/EC: Art 7, 10). Energy certificates are required when buildings were constructed, sold or rented out; certification should be carried out by qualified experts. Buildings larger than 1000 m^2 were to display the energy certificate clearly visible to the public. The intention was that information on the energy performance of buildings would affect price mechanisms in the building market. However, the wording relating to building certification was rather ambiguous and open to interpretation. Because the EPBD was developed by the EU Directorate-General for Energy, in Norway the MPE was responsible for adoption of the directive. Initially, administrators in the ministry ignored the energy certification measure altogether, and hardly gave any weight to energy certification when they wrote and prepared a parliamentary proposition on implementation of the directive (St. prp nr. 79 2003–2004). Not unexpectedly, interviews reveal that the politicians had failed to realize that they had thereby endorsed the introduction of a certificate scheme.

Later the Norwegian Energy Directorate was made responsible for developing the draft of a certificate scheme. Interviews show that they did so quite swiftly, but that the ministry was highly sceptical to such a scheme because it was deemed too expensive; their calculations showed that other kinds of energy-efficiency measures would be more cost-efficient. Interviews indicate that this resistance was rooted in an institutional conflict: the market approach conflicted with the minimizing societal cost logic of the ministry. Interviewees show that civil servants in the ministry saw the Enova approach as a far better way of dealing with energy-efficiency challenges in buildings. At this stage, Norway's neighbouring countries were keenly developing certification schemes, more or less in line with the market logic of the EPBD (see Fuglseth 2009). According to one interviewee this did not promote Norwegian adaptation: instead, 'this taught us how *not* to do it'. Eventually, the Conservative/Liberal government decided to put the certification scheme on ice until after the 2005 elections.

Thus, energy performance of buildings came to be established as a political issue in Norway during these five years. All of the four building energy measures were *initiated* between 2000 and 2005. The MPE facilitated the establishment of Enova, as instructed by the Storting, and it started to develop a certificate scheme for buildings, which was an EU requirement. As to the other two processes, they were initiated almost single-handedly by the Ministry of Local Government and Regional Development and its agencies.

Phase 2, 2005–2010: Incremental Developments in all Directions

In the second period, all the measures developed incrementally, along the path set out in the first period. However, political ambitions increased substantially. The red/green coalition government that took office in 2005 developed a political platform that directly targeted the energy performance of buildings (Soria Moria declaration 2005:59). Low-energy dwellings were to become the standard in the new building code, and the EPBD was to be implemented in the course of 2006. Interviewees agree that these formulations were agreed without much deliberation. The text primarily reflected inputs from environmental organizations and had been rushed through in the shadow of the gas power and CCS struggle. The environmental organizations had many proposals, with energy policy for buildings far from topping the agenda.

In this period, energy efficiency gained normative importance within the European environment. The EU continued on the path mapped out in the early 2000s. In 2007 the Council embraced an indicative target of 20 per cent improvement in energy performance, applicable for the period between 2005 and 2020 (Council of the European Union 2007). However, this objective was not binding for member states, and no specific target was developed for the energy performance of buildings. Energy efficiency was not included when the EU initiated its climate and energy package in 2007, but a revision of EPBD was finalized in 2010. This re-cast followed the same lines as the first directive, but with significantly greater focus on buildings with extremely low energy consumption (Directive 2010/31/EC). Nonetheless, the policy continued to have a 1000 Flowers quality, opening for the use of many national measures and giving the member states significant leeway in their design of these.

More remarkable changes occurred within the EU member states. By 2008, almost all of them had adopted energy-efficiency targets of some kind, although the method of measurement varied so much that actual levels of ambition are hard to compare (Commission 2009:7). Now most member states introduced more holistic energy calculation methodologies in their building codes (EnR 2008:8; Visier et al. 2003:11). A majority introduced state aid schemes, most of which followed technological development logic (Commission 2009:14–15, Jensen 2009:47–50). The development is not quite as impressive in relation to energy certification, as this policy recipe did not diffuse so easily. Despite the requirements in the first EPBD, only a handful of EU member states had introduced energy certification schemes by 2009 (Commission 2009:14; EnR 2008:9). These schemes differed greatly, with some working more as a standardization tool than a market measure (Commission 2009:13–14; Jensen 2009:16). How did this heightened EU focus on the energy performance of buildings affect the Norwegian policy process in the second epoch?

The *Housing Bank* achieved formal political endorsement in the environmental action plan for building construction launched by the Ministry of Local Government and Regional Development in 2005 (KRD 2005:18; St. meld. nr. 23 2003–2004:19). In the coming years, the Housing Bank gradually stepped up its activities, but interviews show that it attracted little political attention (see also KRD 2010). A few major dwelling producers started to offer low-energy buildings as part of their portfolio, and these had on-going dialogue with the Housing Bank on how to improve the sale of these dwellings.[2]

Turning to the *building code* process, we see a shift in pace and ambitions. In 2006, the government launched a draft for energy requirements, with far higher ambitions than the draft that the previous government had put on ice (KRD 2006a). According to interviewees, this reflected the ambitions of the new political executives in the Ministry of Local Government. All the same, most political interviewees admit having devoted little time and energy to the issue. According to one interviewee, greater ambitions in this case were merely the result of 'high personal engagement' among certain political appointees. The deputy minister, Guri Størvold from the Centre (previously Agrarian) Party, played a particularly important role, aiming to realize her party's rather ambitious climate-policy programme.

We do not find the same close relationship between governmental organizations and commercial actors in relation to the building code as with the development of the Housing Bank measures. Dwelling production is an exception to the general picture. Whereas the sub-group of dwelling producers is dominated by a handful of firms that control the whole building chain, the industry in general is split up into a large number of single-purpose organizations. Few firms encompass more than one profession: each group—carpenters and plumbers, for instance—are engaged by different, specialized companies. Manual labour-reliant building construction firms and architect firms employ an average of four people (BNL 2009; Pedersen and Steen 2005; SSB 2009). Construction contractors firms are larger, but they cover only a small share of total building construction (see e.g. NCC 2008, Skanska 2008). Because the constellation of actors shifts from project to project, there are few stable and enduring ties between corporations (Dubois and Gadde 2002). Whereas the producers of single-family houses remain in charge of the whole construction process from beginning to end, the construction of other buildings tends to be organized by tenders and contracting, creating a shifting pattern of intermediate project organizations (Johnstad 1993).

Due to these circumstances, few commercial actors have resources to devote to political activities. True, many have now joined in a common business association, but interviewees underline that the association lacks lobbying resources and has had a hard time developing common position due to internal conflicts (see BNL 2004). It was only after six years of deliberations between governmental experts and researchers that some industry actors raised their voices (KRD 2006b). Inputs to the public consultations show that, in general, the building construction actors accepted the logic of technology development on which the draft for new energy requirements was based (KRD 2006b). However, they called for somewhat lower ambitions. Therefore the level of ambition was reduced from 30 to 25 per cent improvement (compared with the building code from 1997)—which still represented a significant strengthening. The new building code was issued in January 2007, and entered into force in August 2009 (KRD 2007). Later the political ambitions increased: in 2008, a united Storting (except the Progress Party) raised the level of ambitions considerably, agreeing that government should consider the 'passive house' as the standard for all new buildings from 2020 (Innst. S. nr. 145 2007–2008:25).

During the revision of the Norwegian building codes, the European Committee for Standardization (CEN) launched 31 standards relating to EPBD (CEN 2009). Interviewees underline that these were used as a template for developing the new Norwegian energy requirements, although several features were altered in the course of the processes. These are the only European developments that are reported to have influenced the process. The fact that most European countries developed their building codes in the same period as Norway does not appear to have affected the processes much. No actors seem to have referred to EU developments in order to strengthen their position in the national process.

Turning to the *Enova scheme,* we see that Enova had problems getting into dialogue with industry (Riksrevisjonen 2010a:38). Building construction actors did not follow a fully-fledged market logic, in contrast to the stationary energy producers that Enova related to in connection with other aspects of its work. The industry mixed the logic of technology development and the market logic, and had a short-term perspective alien to Enova. Thus, Enova faced considerable problems in working together with actors from the building construction industry. Because the latter were unfamiliar with societal cost minimizing approaches they neither recognized nor understood Enova's own cost-minimizing approach. As one Enova interviewee put it: 'We could not speak economics with the construction actors. They didn't understand us'. And one industry representative remarked: 'It has been very hard. They [Enova] do not understand the construction industry.'

Despite these problems relating to the societal cost minimizing design of the Enova scheme, this approach became stricter and more streamlined over time. This was related to the adoption to the EU regulations on state aid explained in chapter 5. Any use of fiscal measures aimed at promoting buildings with good energy performance must comply with EU regulations. As a main rule, the EU prohibits all state aid (Vedder 2003). The Community Guidelines on environmental state aid specify the exceptions to this main rule (Community Guidelines 2001, 2008). Eligible costs are calculated by an extra cost approach, which entails support corresponding to the additional costs of a building with good energy performance, against a conventional building. State aid is legitimate only if the various kinds of actors that effectuated the same measures received the same level of support. Whether other energy-efficiency efforts would have been more efficient is irrelevant in this respect; what counts is the technological quality of the measure, not how much support is granted. It is the EFTA Surveillance Agency (ESA) that endorses national support schemes, and no measures are to be implemented prior to such notification (Community Guidelines 2001, 2008). If Norway grants state aid in conflict with the rulings of the ESA or the Commission, the recipient must repay it.

Back in 2002, the ESA started to question whether the Enova practice was in line with the regulations on state aid. The Agency focused on the technological criteria applied, not the societal costs. This contrasted with the original design of the Enova scheme, which was designed to support those projects that would yield the most in saved energy per Norwegian kroner in support (Riksrevisjonen 2010a). However, Enova initially also granted support to training and education, without requiring that this should result in measurable energy savings and without ensuring that all applications were granted the same economic advantages.

The MPE defended the scheme by reference to its minimizing societal costs approach, and put considerable effort into ensuring that no applicants received more support than needed to make projects break even. However, the initial concern of the ESA was that some applicants might have been given favourable treatment; the ESA wanted to ensure that actors that offered or applied the same technologies or techniques should be granted the same level of support. The dialogue between the ESA and the MPE continued for four years. Eventually, the ministry was able to persuade the ESA to approve the ministry's cost-minimizing approach.

However, now it dawned on ministry officials that some of Enova's practices were not really aligned to the societal minimizing logic. As noted by one interviewee: 'We actually ended up making things harder for ourselves (. . .) In fact, it did not function as well as we had said.' This was particularly true for some energy-efficiency measures. In 2006, the ESA accepted the Enova scheme, but it also decided that some of the grants for energy-efficiency training purposes were unacceptable (Enova 2007). Thus the recipients had to pay the funds back to the Norwegian state. On this backdrop, Enova halted all support to training and strengthened the cost-efficiency criteria of its scheme even further.

Our empirical investigation indicates that the change in government in 2005 did not impact on how the MPE acted in the ESA process. Interviewees gave various reasons for this. According to one interviewee from the administration, 'they [the political executives] have said that this is nice, but it has not been an important case'; but one political appointee noted: 'this was very difficult, nothing happened, no matter what we did'. Moreover, my interviews show that the politicians never engaged in the dialogue with the ESA. Neither did any actors offer arguments relating to developments elsewhere in Europe: after all, the Enova practice differed from schemes in other countries that were more aligned to the logic of technology development. No one actively used this as an argument, so it did not affect the design of the Enova scheme.

Turning to the *energy certification* processes, also here we see that the Ministry of Petroleum and Energy initially opposed the approach promoted by the EU before finally launching a consultation draft in 2007, after several years of delay. This draft proposed that building owners should ensure certification, not independent experts as prescribed by the EPBD (MPE 2007f). Otherwise the draft was vague and ambiguous. Moreover, it was founded on a different kind of energy calculations, which would mean that even new buildings in line with the new building code could risk getting a low score on their energy certificate.

Back in 2006, when Norway passed the implementation deadline, the ESA started to probe into the Norwegian adaptation of the EPBD (ESA 2008a). Interviews show that due to the large implementation deficit in Europe, the administration in the Ministry of Petroleum and Energy thought that ESA would not pressure Norway for rigid implementation of the EPBD. Thus, no one worried about the slow pace and the divergence with EPBD. Thus, it came as a surprise when the ESA in November 2008 referred Norway to the EFTA Court for infringement of the EPBD[3] (ESA 2008b). The reason cited for the court case was that Norway had failed to meet the implementation deadline. In May 2009, the court supported the ESA claim against Norway (EFTA Court 2009). Brussels interviewees underline that the court case was in part the result of lack of Norwegian engagement. According to one interviewee, 'it was very difficult to get any kind of information! (. . .) It is a mystery to me how the people in the ministry work.' An interviewee from the MPE admits, 'it is obvious that we underestimated the importance of this directive.' One month after the ESA had instigated the court case, the Norwegian government launched a parliamentary bill for a certification scheme (Ot. prp. nr. 24 2008–2009). Interviews confirm the court decision was regarded as an embarrassment, and that the swift process was a direct result of this.

And yet, the energy certification scheme implemented in July 2010 deviated from EPBD requirements, in that dwelling owners were to certify their own buildings (although experts were to ensure the certification of commercial buildings). Further, the certification scheme still followed an energy calculation methodology different from that of the building code. Many building constructors were displeased with these features, but they did not do much to get things changed.

How did all these lengthy processes affect the actual performance of the building construction actors? We know little about how the different measures worked in conjunction, but interviews indicate that the many processes helped to heighten the attention given to energy efficiency by commercial actors. For instance, according to one interviewee: 'the environment was for a long time synonymous with ugly buildings (. . .) but we have noticed a considerable change lately'. Interestingly, it is only among the dwelling producers that a significant number of major actors have started to offer low-energy buildings.[4]

Finally, whereas the activities of the Housing Bank and the building code revision were co-ordinated to some extent, the Enova scheme and the energy certification developed in isolated processes. Interviews indicate that the lack of harmonization was caused by the division of authority; all actors focused on the measures where their own organizations had formal responsibility. Moreover, the Ministry of Petroleum and Energy lacked the technical information relating to building construction that was available to their colleagues in the Ministry of Local Government and Regional Development. Politicians with experience from the MPE recognized that this made it particularly challenging to govern the development of energy policy for buildings, but they did not initiate changes in the lines of division. One interviewee said that the shared responsibility between the ministries was merely 'one rarity you may find among the ministries', and that this was 'hard to change'. The politicians said that organizational change is cumbersome and often entails high political costs. This issue was simply not important enough to spark off new initiatives.

From 2005 to 2010 all four measures continued to develop along their initial path. Neither politicians nor the commercial actors were strongly engaged: governmental administrators served as the drivers. Intervention from the ESA affected the Enova scheme as well as the energy certification, and the EPBD introduced certification. The outcome of the energy policy for buildings was the result of several complex processes. We now turn to the mechanisms that affected the policy emergence and development over time.

6.4 ASSESSMENT

Norway's energy policy for buildings emerged because the politicians called for an energy-efficiency policy, but it was the field-level actors that translated this into meaning a policy that targeted the energy performance of buildings. Political normative steering signals were enabling factors, but the magnitude of policies far outweighs the political instructions. The diverse character of the policy reflects the multi-faced institutional character and the loose couplings of the organizational field; various governmental organizations have had leeway to develop measures in line with their preferences.

Table 6.1   Mechanisms in the case of energy policy for buildings

Phase ⇨ Social system ⇩	Phase 1, 2000–2005	Phase 2, 2005–2010
Political field	*Random Decision-making*	*Random Decision-making*
Organizational field	*Pluralism* Spider Importer Shrewd Lawyer	*Pluralism* Spider
European environment	*1000 Flowers Unpredictable EU Governing*	*1000 Flowers Unpredictable EU Governing*

This variation has been further enhanced by implementation of a range of European impulses: translation of these impulses enabled the governmental actors to enhance their imprint on different strings of policy measures.

Table 6.1 presents the mechanisms at work. Let us now examine the relative importance of the political field, the organizational field and the European environment, and how mechanisms in the three social spheres have influenced each other over time.

The Political Field: Endorsing, Not Steering Policy Development

The political field was dominated by Random Decision-making throughout the decade. Institutionally this issue-area was characterized by garbage can: building construction is a fairly de-politicized area of Norwegian political life, so the political actors have had little to gain from engaging in related issues. Due to lack of politicization, the energy performance of buildings received rather haphazard political treatment; at no stage did the executive government or the majority in parliament given clear and consistent signals as to how the policy measures should be designed. Structurally, resources have been distributed, with the Ministry of Petroleum and Energy responsible for one string of measures and the Ministry for Local and Regional Development for another.

Why did Norwegian energy policy on buildings emerge? The parliamentary decision in 2000 jump-started the policy development. This was a precondition for Housing Bank activity and the creation of the Enova scheme. However, the decision was ambiguous and concerned only with energy efficiency in general, not specifically the energy performance of buildings.

Governmental bodies interpreted this as a promotion of high energy-performing buildings, and that was a necessary precondition for most of the measures to emerge. Through the efforts of Enova, the Housing Bank, and the Building Agency, three different kinds of measures emerged, which were later endorsed by the politicians—generally without much debate. The building code is the exception. Here, the Liberal/Conservative government in 2005 intervened by postponing the process until after the elections. Those elections brought a change of government, however, and the new political executives put the process back on track and introduced higher ambitions. Only one measure lacked explicit political rooting: energy certification. This was clearly not politically initiated, but a result of the European Environment influence, through the EPBD.

To the extent that this was discussed politically, two professional logics set their mark on the deliberations: the logic of societal cost minimizing and the logic of technology development. The governments in the period under scrutiny did not favour either of the two: they simply accepted the proposals they were given, irrespective of the accompanying logic. The cost-minimizing logic gained influence because it was championed by the political leadership of the Ministry of Petroleum and Energy. For instance, the Storting agreed the MPE subsidiary Enova should develop a cost-minimizing fiscal incentive scheme, as proposed by the ministry. The political leadership in the Ministry of Local Government and Regional Development endorsed measures embedded in the technology-development logic, such as the activity of the Housing Bank and the energy requirements of the building code.

Market logic has hardly been debated politically. The certificate scheme is based on market logic, and hence it had a certain market character. However, the administration of the Ministry of Petroleum and Energy did contribute to reduce the societal costs of the measure, thereby lessening its market character. The politicians let the administration deal with policy specification; they did not become involved in shaping how the measure was developed in detail. In general, administrative actors and experts, and to a certain extent industry representatives, were given leeway to develop measures in line with their own preferences.

We do not find any examples of situations in which the political appointees became involved in altering the institutional logic underpinning the various policy measures. Instead, the politicians explain that to the extent they became involved; they were acting on their 'gut feelings'. Political normative ambitions increased in 2005, but the only tangible consequence was that individual politicians engaged in the development of the building code development.

Political engagement was characterized by classic 'garbage can' features: involvement seems to depend on whether politicians had the time, had a specific personal engagement in the issue, or whether the issue was linked to other kinds of climate-policy development. We do not see any example of entrepreneurial activities. True, we see that some politicians engaged more than others, such as Deputy Minister of Local and Regional Development, Guri Størvold. However, it seems that she merely used the structural power she was given through her position, rather than deploying entrepreneurial strategies.

The political endorsement of energy efficiency was a precondition for policy development, but the political field had little independent impact on the character of the policy outcome: the politicians were mainly reactive, simply accepting all policy proposals presented to them. This contributed to the inconsistency and lack of coordination among the measures. The contrasting features of the four measures may prove to be a source of future conflict, but we have not seen much of this as yet. Moving on, we will see how lack of political conflict underpinned layering and incremental policy development, along the four initial paths, at organizational field level.

The Organizational Field: Underpinning Policy Diversity

The organizational field is as diverse as the policy outcome itself, with a broad distribution of structural resources. It consists of a diverse group of governmental organizations, with the Ministry of Local Government and Regional Development, the Housing Bank and the Building Agency as the most the central, while the Ministry of Petroleum and Energy operates on the fringes. Also the commercial actors are a highly diverse group; very few have the resources to participate in strategic discussions of a political nature. Actors regularly involved in building construction relate to each other through chains of weak ties. The exception is dwelling production, where the links are stronger; and here we find rather strong ties to the Housing Bank as well. All in all, this is a prime example of a loosely coupled field.

The field also has a mixed institutional character. There seem to be enduring tensions between the logic of technology development and the market logic, although the former dominates. Most governmental organizations are embedded in technology development logic, but not the MPE and its agencies, which are embedded in cost-minimizing logic. These three logics create three different understandings of the energy performance of buildings: the technology development logic is inclined to embrace a holistic, technologically defined understanding of the total energy performance of buildings, whereas the minimizing societal cost logic leads to a focus on piecemeal improvements of those construction components that are the least costly. Commercial actors embedded in the market logic focus solely on the construction components they deliver, not on the holistic performance of the entire building. Their aim is maximal profit from their particular energy performance-enhancing product, not profit related to construction of high energy-performance buildings as such. The short time-horizon of the commercial actors and their lack of planning capacity make the logics appear in distorted and mixed versions, and this also renders their approaches to energy policy less consistent.

Hence, the field has a pluralist character, with many viewpoints. Relationships between actors are characterized by collaboration and negotiation; no one actor has the structural resources to hierarchically steer the others. Concerning policy initiation, it was the experts of the field—the governmental organizations and the researchers—that played the main roles. Initially, only the Housing Bank and the research community supported improvements in the energy performance of buildings. The Bank was important in the effectuation of the ambiguous political target from 2000 and the transformation of this general energy-efficiency target towards a specific policy directed at buildings. The researchers who worked together with the Bank transferred this perspective into the process of developing a new building code. Here, also the Buildings Agency contributed to promote the new policy development.

Field-level actors played less important roles in relation to the emergence of the Enova scheme and energy certification: the first was initiated politically, whereas the latter resulted from the implementation of EU policy. Commercial organizations failed to appear as either strong promoters or opponents of these new developments: they hardly played any role at all when it came to policy initiation.

Briefly put, then: field-level exerts promoted the emergence of two of the measures, whereas the two others emerged independently of field-level activities. The field-level actors introduced technology development and cost-minimizing approaches in policy measures. On the one hand the Ministry of Local Government and Regional Planning, its agencies and the research communities championed the logic of technology development. On the other hand, the MPE was in the vanguard of societal cost minimizing. Process tracing shows that the former group of actors set its mark on Housing Bank measures and the building code, whereas the latter shaped the Enova scheme. Market-logic commercial actors did not engage much in processes of policy development. True, the energy certification scheme was based on market logic, but because commercial actors were scarcely involved in this, it cannot be ascribed to their efforts.

We see four instances of entrepreneurial activities from organizational field actors. First, technical construction experts in the Housing Bank and the research communities exerted Importer entrepreneurship: they actively brought EU trends into Norway, not only by talking about them and creating Norwegian terms and concepts, but also by building demonstration dwellings that provided the foundation for a Norwegian conceptualization of low-energy buildings. The new European conceptualization of buildings was not forced upon them: they voluntary embraced it and made it their own. The strategy of the experts reflected the technology-development logic in which they were embedded, so they developed measures based on technological, not economic, criteria. The introduction of a Norwegian version of buildings with high energy performance underpinned the expert community's development of the new energy requirements in the building code.

Second, the environmental organizations promoted higher ambitions in relation to the governmental declaration in 2005. This Spider entrepreneurship was based on the Importer entrepreneurship of the Housing Bank and the research communities: the 'low-energy building' concept. The environmental organizations exploited the window of opportunity that appeared when the red/green parties that had won the election started to negotiate the governmental declaration. Initially, the environmental organizations proposed that the EPBD should be implemented by 2006 and that low-energy buildings should be 'standard'. This was an *ad hoc* entrepreneurial activity: none of the environmental groups paid much attention at the time—this was only one out of a many proposals for the negotiations. Because the proposal was regarded as uncontroversial it was readily included in the ensuing 'Soria Moria' governmental declaration. The politicians were clearly not aware of what they had actually decided—that it implied implementation of energy certification of buildings was all in the dark. The loose reference to the EU directive made the decision more ambiguous, easing the political acceptance but also blurring the actual content of the decision.

Third, the MPE exercised Shrewd Lawyer entrepreneurship related to interpretation of EU regulations on state aid; the EU regulations were not entirely clear, but, due to the intervention of the ESA, the ministry and the ESA were forced to perform entrepreneurship. Rather than assessing which arguments would make the Enova scheme appear to be in line with the regulations, the administrative staff in the MPE interpreted the EU regulations on state aid in line with their own professional logic. Hence they argued that the ESA should adapt to their societal cost minimizing practice, rather than the other way round.

As discussed in chapter 5, MPE officials proved to be very good persuaders and succeeded in changing the external EU signal. This was to have a paradoxical consequence: adoption of the EU regulations led to a strengthening of the societal cost logic, even though the regulations were not embedded in this logic. Enova became a victim of the ministry's success and had to ensure that its scheme became even more aligned to the logic of minimizing societal costs. Moreover, Enova was not free to alter its practice when it later became clear that the cost-minimizing logic was an impediment to communication with building construction actors. The entrepreneurship was successful, but it had various unforeseen consequences.

Fourth, the MPE's adoption of energy certification resulted from what was basically half-hearted Shrewd Lawyer entrepreneurship. Initially, the ministry did not understand that they would have to engage in entrepreneurial activity in order to prevent the measure from being introduced. Later, administrative staff in the ministry realized that energy certification would be costly and in contrast to a cost-minimizing logic. Still, they continued to ignore it altogether, apparently without actually planning for realization of the measure. Only after the EFTA court case did the MPE administrative staff acknowledge that creative re-direction of the steering signal was needed in order to circumvent the soft requirements of the EU directive. The staff suddenly found themselves cornered and had to rush through a decision.

The stressful situation that the administrative staff created for themselves further impeded creativity and thoughtful deliberation. They ended up including most of the elements described in the directive, although they did alter the content to a certain extent, as by not requiring independent experts to perform certification. The resultant measure was not aligned to their original logic, nor was it fully in line with the market approach. The EU did not possess strong coercive measures and the EFTA court case merely related to deadline issues and not the actual content of Norway's implementation. Implementation of the measure resulted from a combination of weak structural pressure from the EU and the long time it took before civil servants in the MPE realized that this was an issue that required them to act as entrepreneurs.

We may conclude that the field caused much of the policy diversity; all interested field-level actors were given leeway to develop instruments that fit their internal logic and way of operating, without anyone paying much attention to the interrelationships. Counter to pluralist assumptions, field actors did not engage in a field-wide competition over policy solutions. Instead, the various governmental organizations at field level had leeway to develop the kinds of measures that they favoured, while the commercial actors largely refrained from participating. Without information, or familiarity with the policy development process, or ties to the political field, the commercial actors simply lacked political clout.

Even though the field lacked unity, and was marked by constant crisscrossing conflicts, it exerted significant policy impact. Once initiated, four policies developed along the same path. The main explanation seems to be the lack of clear demarcation lines, combined with the commercial actors' low capacity for policy participation. This meant that potential conflicts were seldom materialized in political debate. The empirical information is less clear on whether the commercial actors did act on the basis of the policies. We know that the activity of the Housing Bank led to changes among the dwelling producers, but otherwise it has been difficult to trace any changes in industry practices. Moreover, with the building code and energy certification, it is as yet too early to assess whether they have been acted upon or not.

European Environment: Escalating the Diversity

The European Environment was characterized by 1000 Flowers: many different measures were adopted at the national levels, the EU was not given much authority, but it adopted a range of measures and recommendations for member states and EEA states. In most European countries, the new attention to buildings evoked the traditional underpinning of building construction policy, the technological logic: the technological criteria in building codes were strengthened and various state-aid schemes based on technological criteria were launched. However, the EU policy, energy certification in particular, was also based on market logics.

European features were brought into the Norwegian policy process in three different situations: the introduction of a holistic perspective on the energy performance of buildings, the energy certification of buildings, and the streamlining of the state aid scheme. Only one measure was not initially promoted by any Norwegian actors: the energy certification of buildings, which was introduced through the EPBD. Social pressure from the European environment for implementation of such a measure was not high: energy certification was not popular in Europe and the number of EPBD infringements was high. Further, the EU had not been granted much competencies or coercive power in this respect, and the content of the directive was rather ambiguous. Rather, it seems that the EU policy had influence on the adoption of energy certification because Norwegian politicians and civil servants alike failed to recognize that they would have to engage in negotiations with the EU, applying shrewd entrepreneurship to get acceptance for a system more in line with Norwegian traditions.

It was primarily governmental experts, not commercial actors, who introduced European elements into the Norwegian debate. The industry has not Europeanized to any great extent, it has not been active in EU policy development and hence does not bring European impulses into the development of national policies. Norwegian research communities and the Housing Bank were more internationally oriented. It was they who brought the holistic understanding of the energy performance of buildings into Norway and based their policy initiatives on this. Moreover, because the MPE was responsible for adoption of state-aid guidelines, they were able to shape how this was interpreted. The governmental organizations were not mere carriers of European impulses: they transformed the European features. And the lengthy process related to EU acceptance of the Enova scheme in relation to the final outcome created both uncertainty and inflexibility.

We see that the European environment helped to introduce one new measure, and contributed to underscore the differences between the Enova scheme on the one hand and the building code and Housing Bank and building code measures on the other. This influence from the European environment came primarily through entrepreneurial mechanisms.

6.5 CONCLUSIONS

This chapter has explored and explained development of the energy policy for buildings in Norway. The policy layering and the 'All in One' character seem to reflect mainly the pluralist nature of the organizational field. Political adoption of an ambiguous target enables various field-level actors to initiate policy development. Political commitment is a precondition—but in this case, policy development benefited from the commitment being rather shallow, as that gave administrative actors the freedom to take political initiatives. Politicians appeared mostly reactive to field-level developments, not as an independent force.

Classical political science contributions on segmentation assume that politicians have little independent importance when the field is highly segmented and institutionalized, and more power when the field is loosely coupled. That was far from the situation here: politicians had relatively little importance even when the organizational field was weak. The activities of others were all the more important; researchers, and administrative staff in ministries and agencies came to dominate as entrepreneurs, while industrial actors and politicians were absent. The skewed pattern in entrepreneurship seems the result of societal background factors, including the low degree of politicization and the lack of political resources among industry actors. Even though the commercial actors in some instances were dissatisfied with the governmental measures they did not engage politically to bring about changes. Thus the policy measures remained stable even though they did not always lead to the intended changes in industry practice.

This case proves the strength of a weak organizational field. We have seen how the nature of the organizational field is likely to have a major impact on the evolution of policy development, also when the field is loosely coupled and ridden with conflicts. In contrast to the assumptions of classic segmentation and neo-institutionalists, lack of unity and agreement at the field level did not affect the clout of field-level actors. Instead, the many different viewpoints on how things should be done served to enhance the development of policy, as the number of measures increased. Had the field been segmented, the policy would probably have developed in a radically different manner. This indicates that the nature of the organizational field will always have a profound impact on policy outcome. If so, then political scientists should always take the nature of the field into account in national policy assessments.

Lack of political steering helped each of the four policies to develop along the initial path, but we should also note that lack of political engagement hampered co-ordination among the measures. With the cat away, the mice did indeed come out to play. This study also reveals a phenomenon that has been accorded scant attention by Europeanization scholars: that European influences can help to empower actors who are given formal responsibility for implementation—governmental administrative officials—while weakening the political executives.

NOTES

1. Enova also collaborated with the Housing Bank in relation to support construction of model projects and R&D-activities (KRD 2009:15; Husbanken2009). The latter activity was aligned to the traditional approach of the Housing Bank and represented a minor proportion of Enova's overall activity.
2. An assessment of the activities of the 10 largest dwelling producers-Mesterhus, Nordbohus, Blik-hus, Block Watne, Norgeshus, Byggmanngruppen, Selvaagbygg, JM Byggholdt, Boligpartner and Systemhus-shows that 4 out of 10 offer some kind of low-energy buildings.

3. The EFTA Court serves the same purpose for the three non-EU member-state participants in the EEA agreement as the European Court of Justice has for EU member states.
4. This conclusion is based on assessment of organizational changes in the 10 largest construction firms (Veidekke, Skanska. AF-Gruppen, NCC Construction, Kruse Smith, Peab Norge, Reinertsen, Backe Gruppen, Heimdal Entrepenør and Spenncon) and the 10 largest dwelling producers Mesterhus, Nordbohus, Blik-hus, Block Watne, Norgeshus, Byggmanngruppen, Selvaagbygg, JM Byggholdt, Systemhus and Boligpartner). The largest construction companies are decided on the basis of their turnover and the largest dwelling producers are calculated in relation to the number of dwellings sold.

Part III

Comparisons and Final Conclusions

7 Comparative Assessment

7.1 INTRODUCTION

The world that confronts climate-policy analysts is confusing: many actors have stakes in policy development, actors tend to change their objectives and arguments over time; and while some areas of climate policy entail great political hurdles, others may develop almost unnoticed by the elected representatives. This book has focused on organizing the conditions of climate policymaking into concepts amenable to systematic assessments. This chapter explores the patterns we find when comparing all the four case studies: CCS, renewable electricity, renewable heating and energy policy for buildings in Norway. The four policy outcomes as well as the development trajectories differ with respect to steering method as well as degree of state steering; and while some policies have developed gradually and incrementally, there has been swift and abrupt change in other areas. On the other hand, we also find some commonalities in the forces shaping policy development. *What, then, can explain the cross-case differences and similarities in policy emergence and change?*

Against this backdrop, the chapter discusses how political fields, organizational fields and the European environment may influence climate policy, offering some answers to the key question in this book: *why do national climate policies emerge and change?* We will discuss why it seems as if politicians play a key role for emergence of climate policies, but have less impact on policy changes over time. Systematic case comparisons show that politicians have been faced with varying steering dilemmas, depending on the institutional and structural character of the organizational fields. Only by requiring a deeper understanding of the functioning of the specific organizational and political fields, and the interrelationship between the two, can we understand why political executives sometimes merely endorse organizational field developments, whereas at other times they engage strongly to change them. This chapter will also discuss how references to EU policy in political discourse do not necessarily reflect the actual power the EU possesses. Rhetorical moves related to interpretations of EU policy may sometimes alter the development of national policies, whereas at other times such activities may have unforeseen consequences.

7.2 POLICY OUTCOMES: DISTINCT DIFFERENCES IN CHARACTER AND CHANGE PATTERNS

Table 7.1 presents the policy outcomes in all cases (based on Table 1.1, in chapter 1). Outcomes differ with respect to steering method as well as degree of state steering. Moreover, they have each followed distinguishable trajectories, and the pace and character of change differ.

The CCS policy started out as a Governmental Industry Development process from the outset, when the Norwegian Pollution Authority applied the country's traditional Pollution Control Act, instructing gas power plants to apply CCS technology. This decision was contested, and it took almost 10 years from the initial decision until the government banned fossil power plants without CCS. In a parallel development, the magnitude of funding the government spent on CCS technology development increased steadily throughout the period. The government clearly signalled that CCS had priority, almost regardless of the costs involved. However, reluctance on part of the industry hindered full-scale CCS from being realized, underpinning a fundamental lack of stability concerning this policy area. In the words of Mahoney and Thelen (2010), CCS exemplifies policy *conversion;* regulations developed for traditional pollution regulation was interpreted in a new way, dissimilar to interpretations of pollution regulations in other European countries. This resulted in the ban on power plants without CCS. Moreover, the initially small funding for CCS research grew into large-scale state aid support devoted to a specific technology.

The renewable electricity policy started out as a fiscal incentives policy, but shifted towards a market instruments when it was decided to join the

Table 7.1 Climate-policy outcomes in the four Norwegian cases

Policy criteria ⇨ State Steering ⇩	Technological	Economic
Indirect state steering	**TECHNOLOGY STANDARDS** BUILDINGS: Building Code	**MARKET INSTRUMENTS** RENEWABLE ELECTRICITY: Green Certificate Scheme
Direct state steering	**GOVERNMENTAL INDUSTRY DEVELOPMENT** CCS: Technology-specific state aid, ban of fossil power plants without CCS BUILDINGS: Housing Bank Measures	**FISCSAL INCENTIVES** RENEWABLE HEATING: Enova State Aid Scheme BUILDINGS: Enova State Aid Scheme

Swedish green certificate scheme. This policy was subject to considerable symbolic decision-making that hinted at a market shift, but the Enova cost-minimizing, fiscal incentive state aid approach was not abandoned until 2009. The green certificate scheme favours the least costly renewable electricity plants in Sweden and Norway, but does not place a cap on the possible profits of developers. It is technology-neutral, and the price of certificates is set by the certificate market. In both schemes, projects are selected on the basis of economic, not technological criteria, but in the green certificate scheme the state governs indirectly by creating a market which in turn produces economic incentives. It is too early to say whether the policy will became stable, changing only incrementally in the future. Renewable electricity is a *displacement* case: introduction of political competition in 2006–2007 led to an abrupt shift towards a new policy character (Mahoney and Thelen 2010).

With renewable heating, we see no abrupt ruptures in policy development over time. This policy measure started out as a fiscal incentives scheme, and still had the same character 10 years later. True, there was some incremental change: as more technology criteria were introduced, the cost-minimizing, fiscal incentive character became slightly less distinctive over time.

The case of energy policy for buildings took on a hybrid 'all-in-one' character, with measures from all the above-mentioned categories. This policy area comprises four different sub-strings of policy, all of which changed over time and without much internal coordination: the State Housing Bank developed governmental industry-development measures; the Ministry of Local Government and Regional Development developed a building code with technology standard characteristics; a state aid scheme with a fiscal incentives character was in operation throughout the period; and lastly, a market instrument—the energy certification scheme for buildings (operated by the Energy Directorate)—was introduced towards the end of the period. *Layering* describes the changes in energy policy for buildings over time: more and more rules and measures are developed within the same issue-areas, seemingly without much coordination (Mahoney and Thelen 2010).

In contrast to the other cases, renewable heating is more an example of non-change than change, at least in term of policy characteristics. Note, however, that this may be the case where we see the most significant, actual on-the-ground changes, in the sense that district heating developments in Norway mushroomed, whereas there were only minor changes in actual low-carbon investments and activities in the three other cases.

We may conclude that there is significant variation among the incremental change cases as regards the nature of the change. Let us now examine the mechanisms that created such different patterns of policy development in the four cases. Here we should note that whereas Mahoney and Thelen (2010) argue that the kind of policy change will reflect the distribution of structural resources in the political field and the character of the policies as such, this book illustrates more complex causal explanations, where several mechanisms work in conjunction.

7.3 COMPARING NATIONAL FIELDS AND
THE EUROPEAN ENVIRONMENT

The Political Field: Most Important When Failing

To what extent and how do political fields explain differences and similarities between the cases? Table 7.2 summarizes the mechanisms at work in the different cases, illustrating significant cross-case variance in kind and number of mechanisms.

The political field had a strong effect on the CCS policy outcome. For almost a full decade the Legislature Governing mechanism was active: there was political competition over the issue and the authority was distributed, shared between two ministries and a strongly engaged Storting. CCS was a key compromise, necessary for the formation of a new government. The organizational field of petroleum put up considerable resistance to the strong political CCS steering, in turn spurring some extraordinarily creative and skilled political entrepreneurship. Such political entrepreneurship significantly increased the impact of the political field, enabling a shift in the information basis on which formal policy CCS decisions were made, undermining the power of the organizational field and altering how EU regulations were interpreted.

The CCS policy outcome was based on input from actors outside the political field, but the incremental strengthening of the policy reflects the functioning of the political field: at no point after year 2000 did the politicians merely adopt arguments provided by the segmented organizational field. Ironically, when it came to actual ensuring construction of a full-scale CCS facility, the politicians seemed rather powerless. No matter how strong their commitment to the issue-area, and no matter how clear steering signals, problems endured.

Long-term dominance of Legislature Governing ensured considerable entrepreneurial activity, significantly enhancing the political field's impact on the policy outcome. The Fashion Queen entrepreneurship of Prime Minister Jens

Table 7.2 Cross-case comparison of political fields

Issue-area ⇨ Mechanism ⇩	CCS	Renewable electricity	Renewable heating	Buildings
Social mechanism	Politicizing– Legislature Governing	Ministerial Governing– Politicizing	Ministerial Governing	Random Decision-making
Entrepreneurial mechanism	Importer Fashion Queen Spider Shrewd Lawyer	–	–	–

Stoltenberg is worth mentioning specifically: in 2007 he launched CCS as the Norwegian 'moon landing' (analogy to the US *Apollo* programme), making it a prestigious project of national importance. And indeed, many eventually came to regard CCS as a crucial, high-profile project for Norway. This served to raise expectations as to the government's achievements in this issue-area, and pressures for stronger state steering and funding increased. Once established, the 'Fashion' game became a mechanism with self-reinforcing power, further spurring the political competition over the issue. All the same, entrepreneurial success was constrained by opposition from the segmented field.

Also with *renewable heating* there was little political engagement. With the structural resources concentrated in one ministry, the elected representatives hardly paid attention to the issue at all. Politicians merely endorsed the incremental changes that occurred at field level, also the introduction of new technological criteria in the support scheme and escalation of project support after 2005. Because no one posed strong objections to the Enova scheme at the organizational field level, the issue was never politicized. This lack of political engagement was a blessing for policy stability and implementation: incremental change and eventual success with respect to industry change and project realization gave the politicians no incentive to demand policy changes; hence the policy remained remarkably stable, with no shift in policy character. Also in this case, the initial target, adopted by politicians in 2000, was a precondition for emergence of the policy in the first place. Not surprisingly, given the Ministerial Governing nature of the social mechanism, we find no instances of political entrepreneurship here.

Renewable electricity was for many years marked by Ministerial Steering, attracting only symbolic attention from the political opposition. At about the same time, the majority in the Storting endorsed the Enova cost-minimizing state aid scheme *and* called for a green certificate scheme. These decisions had mainly symbolic importance: the politicians simply wanted to show that they supported the development of renewable electricity, without necessarily understanding the actual functioning of the two different policy designs. Their decisions reflected the existence of two conflicting logics (market v. minimizing societal costs) at the organizational field level; but it took many years before the mode of the political field shifted to politicization, and politicians took a conscious choice between the two.

Energy policy for buildings developed in radically different fashion: the political field had little independent impact on the character of this policy. Throughout the decade, politicians were mostly reactive, not proactive. Formal decision-making authority was distributed between several ministries, and politicians followed a 'garbage can' logic; hence, political engagement depended on individual politicians, their interests and capacities. By simply accepting all policy proposals presented to them, the political executives endorsed the 'All-in-One' policy outcome, in turn contributing to inconsistency and lack of coordination among the measures. The initial political endorsement of an ambiguous energy efficiency target in year 2000 (in the shadow of the gas power/CCS hurdle) was a precondition for the later policy boom. The level of entrepreneurship was exceptionally low; politicians did not exert entrepreneurship of any kind.

For many years, lack of firm political steering underpinned policy instability with respect to renewable electricity. Eventually, the symbolic political decisions brought highly tangible consequences: when the issue became subject to high-profile political competition, the government finally committed to green certificates. The political field was crucial for the initial adoption of the societal cost minimizing Enova scheme and the later shift in policy, but the politicians were merely responding to field-level inputs—they were not the motor of policy development as we saw in the case of CCS. It seems somewhat surprising, given the competitive nature of the issue, that there were no entrepreneurial activities from the politicians. This may have been due to the lack of extra time and energy, with all the entrepreneurial activity devoted to CCS. Moreover, the issue of renewable electricity figured high on the political agenda for a rather limited time-period.

This cross-case assessment indicates that the political field plays a significant, self-contained role: it is particularly important for the emergence of policy, but also influences policy development over time. Concerning CCS and renewable electricity, we have seen how the politicians played key roles for policy changes; by contrast, with heating and buildings, the politicians were reactive, merely endorsing field-level developments. Intense political entrepreneurship can help to explain why the CCS policy developed incrementally towards more and more Governmental Industry Development. Concerning renewable electricity, the policy displacement happened because the issue finally became politicized: green certificate promoters would not have succeeded if they had failed to spur politicizing.

As expected, the political field is strongest when the structural resources are distributed and political competition drives political action, but the case of CCS shows that even under such conditions it may have limited powers. The case of renewable electricity shows how the intuitional character of the political field may shift rather quickly. This case also indicates that the political field may take a long time to respond to organizational field inputs.

The two cases of renewable energy indicate that Ministerial Governing ('garbage can' logic and concentration of structural resources) is the mechanism where the political field as such has the least influence on policy development. When this mechanism has operated, the actual development of policy has been steered primarily by organizational field actors; the lack of political engagement has moved actual problem solving downwards, into the ministerial bureaucracy or to be resolved by civil servants and corporate actors working together. The dominance of this mechanism may change over time, but the case of renewable electricity shows that it takes considerable energy to politicize an issue.

The political field often responds to developments in the organizational field, but far from all tensions at organizational field-level are picked up by politicians. Political competition fosters political entrepreneurship, while politicians' willingness to try to enhance their impact will be significantly lower in 'garbage can' situations. This study has been conducted in a comparative fashion, but there is also a relationship among the four cases. It seems as if the high political engagement in CCS has tapped the politicians of 'entrepreneurial energy', thereby contributing to a remarkably low level of political entrepreneurship in the other issue-areas. Political engagement varied so much between the cases because one case—CCS—had far greater political prestige than the others.

However, the intense political competition was not exclusively positive for the development of CCS policy: it also served to obstruct dialogue with organizational field actors on how to develop viable solutions that might be accepted by strong field actors as well as the political leaders. Policies spurred by 'garbage can' processes entail more shallow political commitment, but somewhat counter-intuitively we see that this can be positive for policy stabilization and realization: this was the case for renewable heating as well as buildings.

Organizational Fields: Creating the Political Lines of Conflict

To what extent and how do organizational fields explain the differences and similarities between the cases? Table 7.3 shows the organizational field mechanisms in operation in the four cases. In contrast to what we saw in the political fields, there were no shifts in the social mechanisms over time, but considerable variation over time as regards entrepreneurial activity.

The *CCS policy* was embedded in the organizational field of petroleum, a field marked by segmentation, strikingly similar to the classic, theoretical descriptions of segments, iron triangles and policy monopolies (see Baumgartner and Jones [1993] 2009; Egeberg et al. 1978; Hernes 1983:290; Lieberson 1971; Lowi 1969). Structural resources were concentrated in the Ministry of Petroleum and Energy and in Statoil, which are closely aligned, sharing the same market institutional logic. Moreover, segmentation increased over time, due to developments not related to CCS, especially

Table 7.3 Cross-case comparison of organizational fields

Issue-area ⇨ Mechanism ⇩	CCS	Renewable electricity	Renewable heating	Buildings
Social mechanism	Segmented	Turf Battle	Turf Battle/ Collaboration	Pluralism
Entrepreneurial mechanism	Importer Fashion Queen Spider	Importer Spider Shrewd Lawyer		Importer Spider Shrewd Lawyer

the listing of Statoil on the stock exchange and its merger with Hydro. The initial reluctance to act on CCS signals from governmental agencies forced the government to develop its own CCS expertise. Although this eventually underpinned political steering, it was far from enough to undermine the tendency to increased segmentation.

Despite segmentation, this field contained the seeds of the CCS policy outcome. CCS was promoted by marginalized field actors, technological researchers in particular, who based their reasoning on the technology development logic that had dominated the whole field in the 1970s and 1980s. Environmental organizations, operating at the boundaries of the field and with close ties to the political field, mediated the CCS ideas, applying a range of entrepreneurial techniques. The actual entrepreneurial efforts from researchers were rather modest; it was the environmental organizations that ensured that the CCS idea could circumvent the dominant field-level actors and made it into the political field. Moreover, it was the very special situation in the political field—particularly the need for CCS as a compromise in government-making—that made this entrepreneurship successful. The CCS promoters succeeded because they applied multi-field entrepreneurial techniques; the environmental organizations served as mediators, able to plant the idea in the political field.

The effect of the segmentation mechanism was somewhat different from the picture given by classical political science contributions: gradually, the policy outcome came to contrast more and more with the preferences of the organizational field actors, and not the converse. Moreover, the politicians' commitment to CCS grew increasingly strong. It would not be entirely correct to say that this happened despite resistance from the powerful field actors: instead, it seems to have happened *because* of this. The enduring resistance forced the politicians to work hard, again and again, going out of their way in seeking to get their CCS ambitions realized.

Turning to the organizational field of *stationary energy and renewable heating,* we find a very different picture when it comes to how the field influenced policy development. The renewable heating policy was initiated by politicians. The organizational field of stationary energy production was characterized by turf battles, with constant institutional conflicts between the two structurally dominant organizations; the MPE and Statkraft. Yet, this conflict did not have much effect on the development of district heating policy, mainly because Statkraft and other utilities initially did not believe in the economic potential of this technology. Hence, lower-ranking civil servants in the public enterprise Enova (owned by the MPE) and lower-ranking officials in utilities collaborated and adjusted the support scheme incrementally. This led to investments, and eventually Statkraft and other utilities began to believe in the economic profitability of renewable heating. It seems to have been primarily this micro-level coordination that ensured gradual changes in the scheme, not interference by the dominant field actors. The lack of entrepreneurship may seem rather surprising, given the Turf Battle nature of the field; because neither of the dominant field actors was interested in this issue at outset, it was handled by technical experts—entrepreneurship was absent. Hence, the organizational field dealt with the issue in more of a collaborative manner, while the turf-battle mechanisms we would expect under such conditions were hardly in evidence.

Turning to *renewable electricity,* and relating to the same field, we see that the turf-battle mechanism and entrepreneurship played important roles. From 2000 to 2007, there was a continuous battle between the minimizing societal cost logic of the MPE, and the market logic of Statkraft and other dominant utilities. Here we should note that the divisions for petroleum and for stationary energy within the MPE were embedded in different logics: with stationary energy embedded in minimizing societal cost, and the petroleum divisions embedded in market logic (as in the CCS case). As the industry's interest in wind power grew stronger, the conflict became more pronounced. Field-level actors were not able to cope with this conflict themselves; Statkraft soon prompted politicization of the issue, but it took time before these entrepreneurial initiatives yielded results. Because the politicians had adopted conflicting decisions at an early stage, the commercial actors found themselves dependent upon a political solution. The lines of conflict in the political debate largely mirrored the turf battle at field level. Tensions over the design of the electricity support scheme created an enduring uncertainty that obstructed investments and hindered governmental and commercial actors in concentrating on all the technical details involved in getting a particular policy measure to function in practice.

Entrepreneurship was exercised in two different settings: first by Statkraft and other green certificate promoters, who put the green certificate issue on the agenda in the first place; second by civil servants in the MPE, who re-directed the effect of the EU regulations on state aid. Both lines of action served to increase the institutional conflicts at field level, eventually creating a major political hurdle. Wishful thinking in the utilities and the environmental organizations hindered them in investing sufficient resources for creating a shift in policy design in 2006. It is particularly noteworthy that the entrepreneurs failed to realize that the other dominant actor in the organizational field—the MPE—was actively countering their efforts. The initial entrepreneurial acts were preconditions for the later shift to green certificates. However, it was primarily the problems relating to getting ESA acceptance for a hybrid scheme that led the Norwegian government back onto the green certificate track, not the second wave of entrepreneurial activity. The entrepreneurial promotion of green certificates was triggered by the turf battle mechanism at field level, but the entrepreneurship did not yield rapid results. Hence, the field came to play a central role in shaping policy development, explaining the actual character and stability of the policy measures.

Lastly, the case of *energy policy for buildings* reflects the structural and institutional character of the organizational field of building construction:

the pluralist nature of the field resulted in an 'All-in-One' policy outcome. The field itself caused much of the policy diversity: all interested field-level actors were given leeway to develop instruments in line with their internal logic and way of operating, without anyone paying much attention to the interrelationships. In line with pluralist assumptions, many actors initiated policy ideas; but, counter to pluralist assumptions, they did not engage in field-wide competition over policy solutions. Instead, they all had leeway to develop the kinds of measures that they favoured. In line with pluralist theory, there were significant entrepreneurial activities; but this was entrepreneurship exercised by researchers and governmental employees, not commercial actors. Due to lack of information about policy development processes and weak ties to the political field, commercial actors failed to engage much in policy development at al.

We find little coordination among the four different strings of energy policy for building development; once initiated, they all followed their initial path, resulting in layering of measures. The main explanation seems to be the distribution of structural resources, combined with the commercial actors' low capacity for policy participation. This meant that potential conflicts seldom surfaced in political debate. The empirical information is less clear on whether the commercial actors have in fact acted on the basis of the policies. Even though the field lacks unity, and is marked by constant criss-crossing conflicts it has exerted significant policy impact. Hence, it is not necessarily so that a fields need to be segmented in order to impact policy development: rather the question is whether or not the political field is receptive to and responsive to organizational field developments.

We can conclude that the organizational field had less importance for emergence of the policies, but played a major role in shaping the policy characters. In general, the differences in policy character reflect the pattern of professional logics at field level. This is particularly clear in the cases of renewable electricity and energy policy for buildings. Even though the organizational fields were not at their strongest in these cases, the political field modelled their political responses in line with field developments. Concerning CCS, the political field and the organizational field were opposite poles, and the clash between the two systems forced the politicians to work hard in order to try to achieve their political ambitions. The case of renewable heating has shown that the dominant field-level mechanisms will not always be the main driver in all policy processes related to the field.

European Environment: Primarily Influencing through Entrepreneurship

To what extent and how can the European environment explain the differences and similarities between the cases? Table 7.4 shows that the European Environment played into policy development through a range of different mechanisms in three out of four cases.

Table 7.4   Cross-case comparison of European environments

Issue-area ⇨ Mechanism ⇩	CCS	Renewable electricity	Renewable heating	Buildings
Social mechanism	1000 Flowers Unpredictable EU Governing	Bottom-up harmonization Unpredictable EU Governing	1000 Flowers	1000 Flowers Unpredictable EU Governing
Entrepreneurial mechanism	Importer Fashion Queen Shrewd Lawyer	Importer Shrewd Lawyer	–	Importer Shrewd Lawyer

CCS emerged on the Norwegian political scene when CCS had hardly been debated elsewhere in Europe at all. Although the European environment was characterized by 1000 Flowers the whole time, the Norwegian debate had flourished with references to policy development in the EU. Moreover, views on CCS varied considerably, and for a long time it was unclear which EU regulations applied to this policy measure. Other climate issues were far more salient and more contested at the European level. Many Norwegian actors made reference to the European situation, altering the European signals considerably along the way. The EU regulations on state aid applied here, simply because they concerned all areas in which state aid was offered, but no specific state-aid guidelines in relation to CCS existed. In the post-2005 period, this added an Unpredictable EU Governing mechanism to the European influence. At one point it seemed that the EU regulations on state aid posed a threat to the direct engagement of the Norwegian government in the CCS technology endeavour, but, as we have seen, this issue was eventually resolved by shrewd political entrepreneurship.

The European environment influenced the development of CCS policy mainly through entrepreneurship. By creative interpretation, applying Shrewd Lawyer, Importer and Fashion Queen entrepreneurship, politicians were able to overcome resistance from the organizational field. Note, however, that with respect to state-aid rules, entrepreneurship was required: if the politicians had not taken the lead, the civil servants would have been bound to do so, and pressure from civil servants would probably have influenced the policy outcome in another direction.

As regards *renewable heating,* the European environment played no role at all. This seems to have been a great advantage for the stability of this policy area: it helped to hinder entrepreneurial activity and gave lower-level civil servants and corporate actors room to develop a district heating policy and to ensure the construction of new plants and pipelines.

In *renewable electricity,* several European countries were ahead of Norway, having already developed thriving policies for renewable electricity by 2000. The European environment had a bottom-up harmonization character, with feed-in schemes diffusing rapidly in Europe. However, it was the market initiatives from the Commission and the big European utilities that attracted attention in Norway. At the same time as the green certificate idea gained strength in Norway, it was becoming weaker elsewhere in Europe. In contrast to dominant accounts of policy diffusion among neo-institutionalists, Norway was not tempted to adopt a policy that diffused rapidly in its external environment (see e.g. Finnemore and Sikkink 1998; Sahlin and Wedlin 2008).

The green certificate idea came to Norway through the entrepreneurship of Statkraft, later supported by other utilities and environmental organizations. EU regulations on state aid also played into the renewable electricity debate. These were translated by MPE civil servants, resulting in a streamlining of the original cost-efficiency character of Norway's support scheme for renewable electricity. The two lines of European entrepreneurship were incompatible, contributing to field-level conflicts that eventually underpinned the shift from minimizing societal cost state aid to a green certificate scheme.

Concerning the *energy performance of buildings,* the European situation was far more confusing. The European environment was characterized by 1000 Flowers. National practices varied significantly, and the EU was not given much authority, but it adopted a whole range of measures and recommendations for member states and EEA states. Even the core term 'energy performance of buildings' is unclear and subject to a multitude of understandings and interpretations. The EU has developed a significant volume of policy, but the content has remained incoherent, reflecting several different national practices.

European features were brought into the Norwegian policy process in three different situations: the introduction of a holistic perspective on the energy performance of buildings, the energy certification of buildings, and the streamlining of the state aid scheme. Entrepreneurship was central to all these processes. Only one measure was not initially promoted by any Norwegian actors: the energy certification of buildings. The EU policy appears to have had so much influence on the adoption of energy certification because Norwegian politicians and civil servants alike failed to recognize that they would have to engage in negotiations with the EU, and act as shrewd entrepreneurs in order to get acceptance for a system more in line with Norwegian traditions.

Cross-case comparisons show that, on the whole, the European environment had very little direct importance for the significant differences between the Norwegian climate-policy measures. The exception is the energy certification of buildings, where an EU requirement added to the layering of the energy policy for buildings. However, the European environment had indirect importance, by influencing and enabling actors to perform entrepreneurship. Because the European environment did not play directly into the processes of policy development, its effects may appear more random, with the actual influence depending on which actors received the European signals and how they chose to interpret them. However, the influence of the European environment depends to a certain degree on the actual mechanism that dominates within the European environment. Unpredictable EU Governing mechanisms spurred far more national entrepreneurship than the other social mechanisms in the European environment. Otherwise the level and kind of European entrepreneurial activities hinged more on the social state of the national fields than the character of the European environment.

It is also important to keep in mind that we have not assessed any EU Governs situations. Under such conditions, national policy is more likely to be shaped directly by EU steering, and national actors will enjoy less leeway.

The case comparisons have shown the complex relationship between social mechanisms, not only with respect to the European environment but also the two national fields. Whether or not entrepreneurship is exercised will depend in part on the creativity and skills of the actors involved in developing climate policy. However, as we have seen, it will also depend on the social mechanisms at work. In order to understand the relative importance and role of political fields, organizational fields and the European environment, we must recognize how the various policy-shaping mechanisms influence each other, and particularly how entrepreneurship is conditioned by field-specific social mechanisms. The relationship between social and entrepreneurial mechanisms will be discussed below.

7.4 INTERRELATIONSHIPS BETWEEN MECHANISMS

Political Competition Underpins Entrepreneurship

In this section, we examine how the various social mechanisms of the political field seem to influence the amount and effectiveness of entrepreneurship performed by politicians. Table 7.5 summarizes the relationship between the social conditions in the political field and entrepreneurial activities performed by politicians.

CCS is the only case where the political field influenced policy development through social as well as entrepreneurial activities. This case indicates that in Legislature Governing situations, politicians are encouraged to engage in extraordinary ways in order to win political disputes. Politicians will have considerable chances of influencing policy development in such situations, but, as we have seen, their ultimate success will also depend on organizational field factors. It seems that the political field will operate primarily in line with its own internal dynamics in such situations, with less room for actors from outside the field to become leading entrepreneurial actors.

Table 7.5 Entrepreneurial and social mechanisms in national political fields

Social mechanisms ⇨ Entrepreneurship activity ⇩	Legislature Governing	Politicizing	Random Decision-making	Ministerial Governing
Amount	High	High/Low	Low	Low
Effectiveness	High	High	High for non-politicians	Depends on the organizational field
Case	CCS	CCS pre 2000+ Renewable electricity after 2005	Buildings	Renewable heating + Renewable electricity

In contrast, the cases of renewable energy show how Ministerial Governing creates few incentives for performing entrepreneurship. Concentration of resources within a ministry makes it hard for politicians outside that ministry to influence policy development. The parliamentary majority may ask the government to consider changing course—as we saw in the case of renewable electricity—but because the ministry had the authority over the existing scheme, parliamentary involvement did not in itself lead to abrupt changes in policy character.

One reason for the low level of political entrepreneurship in Ministerial Governing situations is that organizational field actors, and not politicians, tend to have the upper hand in such cases. In general, politicians have scant possibilities to reach out and influence field-internal processes in other ways than actually making formal policy decisions. In Ministerial Governing situations, politicians outside the ministry with formal powers have very limited opportunity to influence policy decisions relating to an issue in which they are interested. Of course, they may try to persuade the politicians in power, but they will probably have few opportunities. In order to increase their influence, politicians will have to try to change the mechanism of the political field, from Ministerial Governing to Legislature Governing or Politicizing. Such a shift will significantly enhance their ability to influence policy through entrepreneurship. It does not seem likely that this will happen without some sort of organizational field backing. As long as Ministerial Governing dominates the political fields, the conditions in the organizational field will be more important than the state of the political field for whether or not field-level entrepreneurs will be able to influence policy development.

We see a mixed entrepreneurship pattern in Politicizing situations. Politicizing dominated gas-power discussions leading up to 2000 (before the focus had turned to CCS); here the politicians acted as entrepreneurs—but this did not happen in the case of renewable electricity 10 years later, when this issue became politicized. Hence, case comparison indicates that Politicizing may spur entrepreneurship, but not necessarily; politicians may merely use their regular formal powers to resolve an issue, as they did in the case of renewable electricity.

The case of energy policy for buildings illustrates the conditions for political entrepreneurship created by Random Decision-making: because many politicians had access to decision-making opportunities relating to the issue, they did not have to engage in advanced entrepreneurial activities to order influence policy. To the extent they became involved at all, they merely used their formal position to do so. However, if some politicians had engaged in entrepreneurship in this situation, there is reason to assume that they might get more done than if the political field had been dominated by the other mechanisms. And here is a paradox: while politicians are more likely to succeed with entrepreneurship in 'garbage-can' situations, they are more likely to perform entrepreneurship when political competition dominates. While Random Decision-making makes it unlikely that politicians will act as entrepreneurs, it will probably render the political field highly receptive to entrepreneurship performed by other actors.

On the whole, then, we find that political actors have very limited capacity for entrepreneurship. As a result, very few political issues will gain entrepreneurial attention from politicians. Further, political entrepreneurship patterns can be path-dependent. As long as political competition is prevalent, politicians are not free to delve into new issues: they must concentrate on overcoming the political hurdles that the politicians before them made into important symbolic issues.

Organizational Field Entrepreneurs Target the Political Field

Organizational field actors have been far more active as entrepreneurs than have politicians. Table 7.6 summarizes the relationship between social conditions in the organizational field and entrepreneurial activities performed by field actors.

Segmentation and Pluralism are the two opposing social mechanisms, creating very different conditions for entrepreneurship at field level. Segmentation

Table 7.6   Entrepreneurial and social mechanisms in national organizational fields

Social mechanisms ⇨ Entrepreneurship activity ⇩	Segmentation	Turf Battle	Peaceful Collaboration	Pluralism
Amount	Some	High	Low	High
Effectiveness	Depends on the political field	Depends on the political field	High	High
Cases	CCS	Renewable electricity	Renewable heating	Energy policy for buildings

mechanisms constrain the entrepreneurial opportunities as well as the chances for entrepreneurial success—because one professional logic is so dominant, few actors will be motivated to act as entrepreneurs. Nonetheless, there will always be actors with differing views; and because information and authority are concentrated, actors who challenge the existing policy will have to act as entrepreneurs in order to accomplish anything. Most importantly, the segmented nature of the field means that the entrepreneurs with new ideas will need to target the political field in order to have a chance of being heard. Thus, field-level entrepreneurs must be able to navigate and understand the political field—and that creates a significant barrier. However, the case of CCS shows that marginalized actors in segmented fields can indeed be rather successful: the CCS promoters gained significant influence because they were able to politicize the issue of gas power, later changing into a Legislature Governing mechanism in the political field, promoting a Governmental Industry Development policy on CCS.

Pluralism situations are characterized by institutional disagreement. This seems to motivate actors to perform entrepreneurship. Because structural resources are distributed, there tend to be plenty of decision opportunities to influence. Hence it is not surprising that we find considerable entrepreneurship and considerable entrepreneurial success in the case of energy policy for buildings. This case also shows how the various decision-making situations will probably not be well coordinated, so different entrepreneurs may succeed simultaneously, leading to inconsistent policy developments. Under such conditions, regulatory agencies have the structural power to make policy decisions without much engagement from political executives, and entrepreneurs have scant motivation for approaching the political field. In the case of energy policy for buildings, a broad array of actors performed entrepreneurship; and because the political field was characterized by Random Decision-making, they succeeded without devoting much energy to influencing the dynamics of the political field.

It seems as if Turf Battle situations will be more characterized by conflicts than the case with Pluralism. The existence of significant institutional disagreement seems to motivate more people to act as entrepreneurs; but, because the structural resources are concentrated, there will be fewer chances for entrepreneurs to be able to influence actual, field-level decision-making. The case of renewable electricity shows how entrepreneurs may depend on politicians to deal with the issue, in order to resolve turf battles at field level, and many years may pass from when the issue is put on the political agenda and until it is resolved.

We find no proper Peaceful Collaboration situations in our case studies, but the case of renewable energy is characterized similar dynamics: due the lack of institutional conflict here, we do not see any field-level entrepreneurship. As noted, this is largely because the issue was handled at a lower level than where the turf battles of the field tend to unfold. No one rose to the occasion to promote an alternative professional logic to the cost-minimizing logic of the Enova support scheme.

All in all, it seems that organizational field actors will perform more entrepreneurship than politicians. However, in Segmentation and Turf Battle situations, the effect of entrepreneurship will depend mainly on the situation in the political fields. The political field has the ultimate authority in most democratic societies, so actors who lack the necessary resources to influence decision-making at the field level will have to approach the politicians in order to be heard. Hence it is not surprising that our case studies show that field-level entrepreneurs have targeted the political field when the structural resources were concentrated at field level, in Segmentation and Turf Battle situations. Organizational field-level entrepreneurs may gain significant political support, even when their ideas and viewpoints go counter to the interests of the dominant actors in segmented fields.

EU Creates Entrepreneurial Opportunities

Our case studies have shown that the Europeanization of climate policy has helped to render national climate policymaking more complex, thereby often providing new opportunities for entrepreneurship. How and to what extent the European environment underpins entrepreneurial activity varies from one issue-area to another, however. National actors have performed entrepreneurship in relation to implementing or translating European steering signals, as well as in relation to other, softer, impulses. Table 7.7 summarizes the relationship between social conditions in the European environment and entrepreneurial activities performed by national actors.

EU Governing and 1000 Flowers are two contrasting situations with respect to institutional unity and distribution of structural resources. In this book we have not examined any EU Governing situations, but it seems reasonable to expect that there will be little room for national entrepreneurship in such situations. With most decisions made in EU forums, there will be less leeway for national actors to influence how the EU policy is implemented nationally. Of course, national actors can influence the development of EU policy by participating in EU decision-making. EU Governing accords little leeway to national actors *after* the EU policy is adopted. Nonetheless, as long as there are legally binding EU rules, there will always be some room for national 'shrewd lawyers' to negotiate over the details relating to national implementation practices. In these situations, the amount of entrepreneurship will be influenced mainly by the situations in the national fields, but the chances of success will be influenced mainly by the European environment. As EU regulation is rather tight and one policy model dominates, there will not be much chance of major entrepreneurial achievements. The EU ETS might fit these expectations—but this should be investigated empirically. From our case studies we have seen how, in 1000 Flowers situations, the EU has very little authority over national policymaking, and the many different

Table 7.7   Entrepreneurial and social mechanisms relating to the European environment

Social mechanisms ⇨ Entrepreneurship activity ⇩	EU governing	Unpredictable EU governing	Bottom–up harmonization	1000 Flowers
Amount	Some (?)	High	Depends on the national fields	Depends on the national fields
Effectiveness	Low (?)	Depends on the national fields as well as the EU legislation	Depends on the national fields	Depends on the national fields
Cases	–	CCS Renewable electricity Energy policy for buildings	Renewable Electricity	CCS Buildings

practices and policy ideas in different parts of Europe. National actors may be inspired by what happens elsewhere in the EU, but, with practices there varying so widely, it will be up to the national actors to decide whether to use European arguments in developing their own policies. This happened with respect to CCS as well as energy policy for buildings. Highly creative, institutionally skilled entrepreneurs succeeded in arguing that Europe had solutions with respect to buildings, and that Norway should lead the way for general development of CCS within the EU. The entrepreneurs targeted national decision-making processes, so their success depended mainly on the situation in these national fields, not the European situation. As ultimate authority rests with the popularly elected representatives, the political field was particularly important.

We have seen that Unpredictable EU Governing creates highly complex situations with respect to entrepreneurship. This happened in the cases of CCS, renewable electricity and energy policy for buildings, where the unclear content of the EU's formal regulations forced national actors—civil servants in Norwegian ministries and agencies—to act as entrepreneurs. In such situations there will tend to be discrepancies between EU legislation and national practices, but also considerable uncertainty as to the size and importance of these discrepancies. Lack of common EU practices also makes it hard for Brussels to take final decisions as to whether a country is complying with EU law or not. Due to the varying institutional features in the European environment, a patchwork of different practices is endorsed, and the rulings of the Commission are rather unpredictable. This leads to negotiations—and here, skilled bargaining actors can have considerable room for influence. Ultimately, however, the strategy of national entrepreneurs will depend on the conditions in the national fields, and some EU rules can be bent more easily than others. Entrepreneurship will succeed only if the entrepreneurs are good at selling their arguments in Brussels. Also the European Court of Justice may contribute to the development of EU practices and decision-making in Unpredictable EU Governing situations, but that is not explored empirically in this book.

The case of renewable electricity shows how the Natural Harmonization mechanism creates considerable opportunities for national institutional entrepreneurship: national actors may refer to what is going on elsewhere in the EU and perform Importer entrepreneurship. Also Fashion Queen entrepreneurship is possible in Natural Harmonization situations, even though that did not happen in the case of renewable electricity. In this case, the Norwegian entrepreneurs operated with a highly creative interpretation of what was happening in the European environment: while feed-in schemes were diffusing and becoming increasingly salient in Europe, Norwegian entrepreneurs portrayed green certificates as the most fashionable and popular European support scheme. Eventually, this line of argument had high impact on the development of Norwegian policy. Ultimately, whether anyone will perform entrepreneurship in Natural Harmonization situations, and whether that will succeed, will depend on the situation in the national fields.

This means that, with the exception of *EU Governs* situations, the case studies indicates that the European environment has the possibility of creating numerous entrepreneurial opportunities, but will rarely influence the chances of entrepreneurial success. In some instances, entrepreneurs will respond to structural features in the European environment (such as binding regulations and directive requirements), but it will often be the entrepreneur who initiates the coupling between the national policy process and the European environment. Note that the authority of the EU may change over time. The more of the policymaking that is done in Brussels, the more will the European environment influence the entrepreneurship chances of national actors. And the strengthened EU climate policies decided in the 2008–2010 period had only to a limited extent started to influence our four cases when this empirical study concluded in 2010. Moreover, entrepreneurs may contribute to strengthen the ties between the development of national and European policies, as shown by the case of CCS.

The chances of succeeding with European entrepreneurial strategies seem influenced more by the situation in the national fields than by the European environment. The latter has been an important source of entrepreneurial inspiration, but only rarely has it had much influence on the entrepreneurs' ability to succeed.

Interdependencies between Political and Organizational Fields

Our entrepreneurship assessment indicates certain interdependencies between the two national fields: the social as well as the entrepreneurial mechanism of one of the fields tends to affect the functioning of the other. There is, however, an important difference: the political field is far more responsive to organizational field signals than the reverse; the political field is to a high degree influenced by the mechanism of the organizational field.

Political fields seems to be responsive to the organizational field in the sense that the mechanisms of the former may change quite readily in response to developments in the latter: for instance, with respect to CCS as well as renewable electricity, the mechanisms of the political fields changed as a consequence of field-level entrepreneurship. And with energy policy for buildings, the political field did not merely adopt the proposals from the field: the political field also took on a similar character as the field— distribution of structural resources and a garbage can logic opening up for political decisions in line with several professional logics.

The two cases of renewable energy indicate that conflicts at field level will influence social mechanisms in the political field only if entrepreneurs actively bring them up for political resolution. The politicizing of the political field in relation to renewable electricity was clearly the doing of organizational field actors, bringing the turf battle from the field into the political field. When renewable electricity was politicized, the line of conflict in the political field eventually mirrored the line of conflict in the organizational field. The character of the organizational field was stable, whereas the mode of the political field was changeable, although it took more to effectuate the political competition than the field-level entrepreneurs had envisaged.

The picture is different when it comes to political fields influencing the mechanisms of the organizational fields. A main reason seems to be that the social mechanism of the organizational field is influenced by policymaking in a whole range of issue-areas, making it more resilient to change. For instance, at the same time as the CCS case was evolving, the political field endorsed privatization of Statoil, and the merger between Statoil and Hydro. These decisions strengthened the segmentation processes in a way that countered the destabilization effect of the political decisions relating to CCS.

The CCS case has a *sui generis* quality, not the least because the strong modes of the two national fields underpinned such a wide range of entrepreneurial acts. Here the political field was at the peak of its powers, but its impact was limited by the equally potent organizational field. We can say that Norway's CCS policy developed through a *Clash of Giants*: both the field and the political field were in their most powerful modes (Legislature Governing and Segmentation respectively), but conflicting mechanisms in the two fields produced stalemate. Because neither of them yielded, the conflict ran into a deadlock, and a solution with a 'saving face' quality for the politicians seemed out of sight.

That was far from the situation in the other issue-areas, and helps to explain why almost all entrepreneurship opportunities were drawn on in the CCS case, but not in the others. For instance, it is probably no accident that we find Fashion Queen entrepreneurship only in the CCS case. Green certificate promoters found themselves in a situation open to Fashion Queen entrepreneurship in the period 2004–2006, but they did not exploit the opportunity. At this point, Sweden and Norway were in the final phase of negotiating the development of a common green certificate scheme. European Commission officials as well as the big European utilities expected Norway and Sweden to lead the way by developing a bilateral scheme that could turn into an integrated all-encompassing European scheme. The Norwegian utilities wrongly assumed that victory was already assured and failed to seize the opportunity for Fashion Queen rhetoric. Perhaps the Norwegian green certificate scheme could have been introduced at an earlier stage if the utilities had seized this opportunity.

In general, the political field plays the key role in shaping whether entrepreneurship will prove effective and succeed. Because the ultimate authority in democratic societies rests with popularly elected representatives, that is probably not so surprising. The more authority that rests with the EU in an issue-area, the more likely are the chances of entrepreneurial success to be determined by the European environment situation. In Unpredictable EU Governing situations, the chances of entrepreneurial success are substantial—but these are highly demanding, complex situations. Hence, it is mainly the most resourceful national actors, and actors involved in European as well as national processes, that will succeed in influencing the decision-making. Most often, this means that Europeanization give civil servants greater influence on policymaking. Moreover, developing good entrepreneurial strategies requires in-depth understanding of the relative importance of national fields and EU legislation.

On the whole, our case comparison shows the asymmetric interdependencies between political and organizational fields. In most instances, the organizational field situation will affect how the political field handles an issue. All the same, the politicians involved in a certain political field can influence the mechanisms that operate at field level only if they succeed in influencing political field decision-making: they cannot perform softer kinds of 'downwards' entrepreneurship.

8 Theory Conclusions

8.1 INTRODUCTION

This book has brought together a broad range of social science insights and combined them in a holistic framework that enables us to understand new aspects of policy development in general and climate policy in particular. Chapter 7 systematically compared and explained the cross-case differences and, on this basis, discussed why national climate policies emerge and change. This chapter will discuss the relevance of the case study results to some on-going discussions in social science.

Particular attention will be given to how the multi-field framework may add to our knowledge concerning the relative importance of political steering, the conditions under which business interest will have greatest influence on policy development, and the importance of entrepreneurship in relation to the implementation of EU policy. This chapter calls for political scientists as well as sociologists to have more realistic perspectives as to how and to what extent politicians may steer developments in climate policy. We will also discuss the paradoxes of entrepreneurship in policy development processes, arguing that sociological neo-institutionalism and field theories as well as policy studies would profit from analytically separating assessment of the volume of entrepreneurship from assessment of entrepreneurial success and effectiveness. Moreover, it does not seem that the existence of many entrepreneurial opportunities necessarily leads to extensive entrepreneurial activity.

Important future research challenges for the multi-field framework will be discussed towards the end of this chapter. This book presents a new agenda for national climate policy studies—as a starting point, not an endpoint for discussions. More empirical studies are needed and many puzzles deserve more attention from scholars of climate policy.

8.2 THE POWER OF POLITICS

Initially, we asked: to what extent and how are the emergence and change of climate policies influenced by politicians and national political fields? The case-study findings indicate that political fields are particularly important for policy emergence, whereas the influence on policy change varies significantly. We find that political steering has played a surprisingly different role in the four issue-areas studied here: politicians sometimes had to invest a lot in order to spur policy development, whereas at other times even the weakest political signals had profound policy consequences. Enduring political attention was *not* a precondition for politicians to influence a given policy development: sometimes also shallow commitment may yield significant effects. This indicates that political scientists would be well advised to stop downplaying the independent importance of political steering (see Hooghe and Marks 2001; Moravcsik 1998; Skocpol 1985). Moreover, our four cases have shown how the power of politics may differ considerably from one issue-area to another.

Political scientists who do pay interest to political steering often turn to 'political will' as an ultimate explanation for political developments. For instance, several environmental policy scholars have cited 'lack of political will' as a superior explanation for low levels of achievement in environmental policy (see Delmas and Young 2009; Miles et al. 2002). Our case studies have shown that 'political will' is a far more complex phenomenon than these scholars seem to take into account. Political engagement is seldom a mere result of normative support for a certain policy choice. True, politicians will often portray disagreements as conflicts over values—but this is frequently a simplification of the actual conflict, which centres on *how* things should be done. For instance, politicians are often faced with the challenge of choosing between measures that are based on different professional logics, but they seldom have the time to assess actual differences between the various measures.

The case studies presented here show that politicians may have rather simple normative positions initially—for instance, being against gas power but in favour of renewable energy. If an issue is characterized by what we have called 'garbage can' logics (as indeed most cases generally are), politicians will probably not have the time or resources to study and understand the differences between different policy designs. If they fail to navigate smoothly among the various policy-design proposals, they may impede realization of their normative objective. We saw this in the case of the renewable electricity, where it took a long time from when the idea of green certificates was born, until the actual political adoption was made.

Differences in the social character and functioning of political as well as the organizational fields will lead politicians to deal differently with different issues, so more political energy will be required to steer some kinds of policy development than others. Whether or not political leaders will engage strongly seems largely dependent on whether the issue becomes subject to political competition. Even if they engage strongly, they may still not succeed. For instance, no one can say that Norwegian politicians have not tried to create a policy that fosters full-scale CCS is Norway. They showed exceptionally high and enduring political will—but they failed, due to resistance from the political field. On the other hand, politicians may also be heavily normatively engaged in issues characterized by 'garbage can' logic; but few will have time and energy to follow these policy issues closely. Further, disagreement is a precondition for political competition to emerge in the first place. This has the paradoxical effect that politicians will pay far less attention to issues on which they all agree, than to the issues where they disagree.

Our case studies also indicate that if there is concentration of structural resources in the political field—with political authority concentrated in one ministry, rather than in the parliament—the issue is likely to evade political steering altogether. With no one watching them, the politicians in charge of that ministry will have less, not more, interest in paying attention to the issue. Due to lack of political engagement, actual problem-solving gets moved downwards, into the ministerial bureaucracy or between civil servants and corporate actors. On the other hand, if the political executives do have an interest in the matter, they will be in a very powerful situation. Things are radically different if authority and information are spread within the political field: the more political actors share responsibility for an issue, the more information will the politicians have, with greater incentives for engaging and giving priority to it.

In line with the arguments of Fran K. Baumgartner and Brian D. Jones ([1984] 2011) and John Kingdon (1984), we find that neither the distribution of structural resources nor the political logic of the political field is stable. Baumgartner and Jones argue that policy change will rarely happen if political actors fail to give much attention to an issue: 'Long periods of stability are interrupted by bursts of frenetic policy activity' ([1993] 2009:xvii). This book has gone deeper in conceptualizing the processes that may destabilize a policy area, and the relative importance of the political and organizational field in this respect. Baumgartner and Jones pay considerable attention to how external shocks may alter the saliency of an issue. Our comparative case studies show that also enduring entrepreneurial work from field-level entrepreneurs can result in an issue becoming politicized. However, politicians can also make important political decisions with deep ramifications for policy development also in non-salient issues. For instance, the boom in energy policy for buildings measures was initiated politically, but the number of measures soon increased far more than the politicians had foreseen.

Like Baumgartner and Jones, Pepper Culpepper (2011) argues that the higher political salience, the more powerful the politicians will be. To a certain extent, our comparative case studies confirm this: the political field *does* tend to affect policy development the most when it is characterized by political competition, particularly when the structural resources are distributed and the Legislature Governing mechanism is in operation. After all, Norwegian politicians managed to get a CCS policy adopted, against the will of a powerful industry. On the other hand, they did not succeed in realizing their CCS objective, despite their strong engagement. We have also seen how political decisions characterized by 'garbage can' logics and shallow political commitment had major ramifications: for instance, it underpinned the boom in measures concerning energy policy for buildings. Our empirical studies also underpins Culpepper (2011:180), who argues that it is a weakness in current understandings of politics to assume that all important decisions about institutional rules are made in legislative arenas. Often, and particularly in areas characterized by 'garbage can' logics, the decisions are made in ministries—prepared by civil servants and then merely endorsed by the political appointees.

Jim March and Johan P. Olsen (1983; see also Olsen 1983) argue that politicians will have more importance for shaping policy agendas than in shaping actual policy development, but they have a bleak view on the importance of political steering. They underline how politicians, more often than other actors, are subject to major time constraints. In many cases, 'a single individual has neither the cognitive capacity, nor the time and energy, nor the moral and representational standing' to independently shape a particular outcome (March and Olsen 1983:283). It is more important for political leaders to state their intent and willingness to try, than to show enduring commitment to a particular political project (Olsen 1983). Further, political leaders may make only modest use of the authority they have fought to achieve, 'because the symbolic value of the authority is more crucial to them than its exercise' (March and Olsen 1983:291). Politicians will at most be gardeners who undertake small adjustments, not engineers who launch major policy changes (292).

Our comparative case studies indicate that politicians are indeed more important than gardeners, but they are not engineers who, unhindered, can design and steer societal developments. The multi-field approach enables us to see that politicians should rather be compared to landscape architects. Just as landscape architects draw and plan new parks and green areas, politicians initiate and supervise new policy developments. In order to succeed, they need to understand the larger landscape in which 'the park' is to be situated; moreover they will have to take into account the competences and inclinations of the 'landscapers and gardeners' who are to effectuate their architectural drawings. In most cases, politicians will give steering signals similar to those they have given before—to continue the metaphor, they will draw parks rather similar to those that landscapers and gardeners are used to construct. This makes the result rather predictable: the final park does resemble the initial drawing, because of the work of the landscape architect, but landscapers, gardeners, soil quality and climate conditions all play into the equation. Political entrepreneurship is not really essential in such instances: the policy development follows kind of a routine.

But sometimes landscape architects have brand-new ideas that require the use of techniques unfamiliar to landscapers and gardeners, or techniques that do not suit the original climate and soil. This requires stronger political steering and more entrepreneurial activity. CCS was one such instance. Because the politicians aimed for a 'park' that required considerable extra schooling, fertilizing and irrigation, policy effectuation could not be delegated to actors at the field level. The politicians had to stretch their powers. The 'gardeners' and 'landscapers' were annoyed at this irregular heavy political engagement, but the entrepreneurship eventually enhanced the politicians' access to knowledge as well as to coercive tools. This resulted in a quite unusual 'park', very unlike the ones the landscapers and gardeners usually constructed, but it was also quite different from the initial drawing.

Policies spurred by 'garbage can' processes entail more shallow political commitment—but, somewhat counter-intuitively, we find that this can be positive for stabilization and realization of the policies. In the landscape architect analogy, this means that great parks and cultivated landscape may sometimes emerge without any landscape architect at all: they may result from the activities of landowners, landscapers and gardeners that contribute to change the scenery without being part of any larger, coherent plan.

8.3 BUSINESS POWER

To what extent and how do climate-policy emergence and change tend to be influenced by national organizational fields? And when will business actors influence the policy the most? Back in 1977, Charles Lindblom argued that business would always have a privileged position in policymaking. This argument implies that business will have a dominant position in organizational fields, and that political decision-making will primarily reflect organizational field-level conditions. Various political scientists have called for their colleagues to pay more attention to how and to what extent business influence policy. According to David Coen et al. (2010: 9), the relation between business, government and policy development 'has long been a stepchild within the [political science] discipline', and this line of research is 'undersupplied with theory'. Arguing that 'the study of business power is currently more neglected than it has been for the last half century', Culpepper (2011: 185–86) writes of 'a neglect of the mechanisms by which business converts its interests into policies'.

The multi-field approach underpins nuanced assessments about how and through which mechanisms business, civil servants and environmental organizations influence policy emergence and change. Our case comparisons have shown that organizational fields are salient to policymaking, irrespective of which mechanisms operate at organizational field level. The organizational field had less importance for emergence of the policies, but they played a major role in shaping the policy characters. In general: the differences in policy character reflect the pattern of professional logics at field level.

We find significant variation as to the relative importance of business actors across fields. Business played a key role in the CCS and renewable electricity cases, whereas civil servants were more important to renewable heating and were central to the development of energy policy for buildings.

However, business did not easily get things its way, neither with respect to CCS or renewable electricity. There is nothing to indicate that business will *always* have a privileged position in policymaking.

The conceptualization of organizational fields enables us to see that in some instances, business will influence policymaking in much the same way as the segmentation scholars have argued, whereas at other times the policy influence of business will resemble the arguments of neo-pluralists. We also see that the politicians can be surprisingly powerful in segmentation situations and that business interests can be surprisingly weak in pluralist situations.

Scholars have used various labels to describe the organizational field mechanism that I have called segmentation, such as 'corporatism' (Panitch 1980; Schmitter 1974), 'interest-group liberalism' (Lowi 1969), 'segmentation' (Egeberg et al. 1978; Hernes 1983:290; Lieberson 1971) 'iron triangles' (Hernes 1983:291) and 'policy monopolies' (Baumgartner and Jones [1993] 2009:7). All these approaches argue that business will have a major say on policy development in segmentation situations. Indeed, the CCS case has shown that the segmented Norwegian organizational field of petroleum was able to resist considerable political pressure concerning the development of full-scale CCS. True, it is hard to distinguish the relative power of the civil servants and corporate actors in this case, because they shared the same viewpoints. However, it seems clear that the corporate actors had the upper hand: after all, it was the administrative staff that adopted the logic of the corporate actors, and not the other way around. On the other hand, this case also shows that, even in segmented situations, politicians can have the power to counter the will of a strong industry, and adopt a policy that goes counter to those interests.

Also pluralists are only partly correct in their arguments concerning business power. With respect to energy policy for buildings, the field level was characterized by highly decentralized, fluid and situational patterns of policy influence (see Lowi 1964:679). Many field-level actors had the opportunity to influence policy development. However, we have not found that mobilization of one group automatically leads to the counter-mobilization of another (see Baumgartner and Jones [1993] 2009:4–5). Neither have we seen much elaborate networking activities on the part of business actors, as described by many central pluralist scholars (see Heclo 1978; Rhodes 1997; Sabatier and Jenkins-Smith 1993a). It was civil service actors and entrepreneurs that had the greatest influence, whereas few industry actors had the resources or time for engaging in policy development.

Neither the segmentation nor the pluralist approach captures the asymmetric interdependencies between political and organizational fields. Generally, the organizational field situation will impact on how the political field handle an issue; however, the political field is far more responsive to organizational field signals than the reverse. The social character of the political field is significantly influenced by the mechanism of the organizational field, and the level and kind of entrepreneurship exercised by field-level actors. Further, the mechanism that drives policymaking in the political field may change quite readily in response to organizational field developments, whereas organizational fields tend to be more inert, changing more slowly.

Distinguishing between the organizational field and the political field enables us to assess the relative importance of the two and their interrelationship. Conventional social science approaches tend to talk about the 'state' or the 'political-administrative apparatus' without distinguishing between the political field and the administration realm (see Christensen and Lægreid 2001; Christensen et al. 2009; Fligstein 2008; Skocpol 1985). This makes it hard to grasp how dissonance and interdependencies between organizational and political fields can be important to policy development. By separating the two, we can assess how political executives are constrained and enabled by the governmental apparatus they govern. It may make sense to conflate administrative bodies and political elects in, say, the US system, where a broad range of administrative leaders are politically elected, but this hinders analytical clarity in a European system where only ministers and deputy ministers change as a result of shifts in government.

An important reason why neither segmentation nor pluralist arguments reveal the full picture when it comes to business influence on policy development is that they fail to take the conditions in the political field into account. In his *Quiet Politics and Business Power,* Culpepper (2011) argues that the salience of the issue in the political field is the key to understanding business power over policy development. In issue-areas characterized by political competition, business will have a rather week position, whereas they will have more impact on 'quiet politics'; in the 'arena of noisy politics, organized business actually suffers many defeats, because these are the conditions under which politicians must cater to popular opinion if they want to be re-elected' (Culpepper 2011:xvi) and 'business power goes down as political salience goes up' (2011:177). Culpepper draws his general conclusions from a thorough comparative case study of national corporate control regulations and the influence of managerial preferences cross-nationally—a completely different issue-area from the climate policies that we have explored.

In one respect, the present book lends support Culpepper's argument: In the most 'noisy' policy area, CCS, the politicians have come a long way in overruling powerful business interests. True, corporate actors were able to significantly impede the ambitiousness of the politicians; but, because of the high political engagement, the political field played a more important role here than in other areas of climate policy. However, political salience is not in itself the explanation to the constrained business influence in this case. At least for the period after 2005, it was precisely the stark opposition from business interests that made CCS a salient policy area. Also with respect to renewable electricity, industry influenced the salience of the issue: entrepreneurship from corporate actors led to politicizing of the support scheme, and eventually the business actors got it their way.

Energy policy for buildings is a clear example of what Culpepper calls 'quiet politics': not once in the years between 2000 and 2010 did the issue gain significant political attention. All the same, this issue-area has been very little influenced by building construction corporations: it is civil servants that were the main driving forces. Even though it in theory could have been easy for corporate actors to gain influence, they lacked the necessary resources for strong political engagement. Thus, Culpepper is correct in arguing that lack of political competition reduces the clout of the politicians, but wrong in holding that this automatically increases the power of business actors. Even if business actors have clear economic interests in an issue-area, they will *not* automatically engage and seek political influence over that issue-area.

Our comparative case studies have shown that the relative importance of political steering decreases when 'garbage can' logics dominate in the political field, but it seems to be the structural and institutional characteristics of the organizational field that shape whether corporate interests or other actors will capture the room for political influence. Dominance of one or a small group of large, professional corporate actors appears to be a precondition for high business impact. Culpepper (2011) argues that it is important to take institutional features into account in order to understand business power—but fails to recognize that institutional unity between civil servants and corporate actors, as the CCS case, can make political steering very challenging.

This book pays greater attention to institutional features than do traditional political science approaches. The multi-field approach shows how the power of corporate actors is *not* a mere reflection of structural resources but also a result of institutional features. For instance, one main reason why there was far more room for political steering in the case of renewable electricity than with CCS was the split in institutional logic between ministry officials and Statkraft in the former case, and unity between the ministry and Statoil in the latter.

The entrepreneurial activity of business actors is an x-factor that may have significant impacts on the ability of industry to influence policy processes. It seems as if Europeanization of climate policy strengthens business actors and civil servants somewhat more than environmentalists, as they have more resources for following EU discussions.

This leads to three conclusions concerning the power of business in policymaking. First, the structural and the institutional character of the organizational fields influence the ability of business actors to influence policy: business seems to be the most powerful in segmented fields dominated by the institutional logic of the business actors. Second, business communities dominated by many small and loosely coupled companies will have less political clout than those dominated by a few large, professional corporations. Third, business actors may significantly increase their influence by entrepreneurship, but they do not appear to be particularly more skilled entrepreneurs than other actors, such as civil servants and environmental organizations. Fourth, marginalized and small actors with few resources may reduce the power of dominant business actors by politicizing an issue and persuading the politicians to take on their cause.

8.4 EUROPEANIZATION AND ENTREPRENEURSHIP

Despite the steep growth in international policymaking, there has been surprisingly little dialogue between students of international politics and public policy on how this alters the power of different actors in national political processes. This is a particularly pressing concern in EU studies: EU policies cover an increasing number of issue-areas, and the authority of the EU over national policymaking is on the rise (Cowles and Risse 2001:218; Olsen 2006:96). The multi-field case studies indicate that examination of entrepreneurship is crucial for finding good answers to the question: to what extent and how do the emergence and changes in climate policies tend to be influenced by EU policy and the European Environment?

European environments had a certain influence on both emergence and change of Norwegian climate policies, but this influence was often indirect, subtle and substantially altered by actors using entrepreneurial techniques to ensure that EU policy and European developments strengthened their political clout. Unpredictable EU Governing mechanisms spurred far more national entrepreneurship than the other social mechanisms in the European environment, but note that the lack of EU Governs mechanisms in the case studies may contribute to create somewhat skewed conclusions in this respect.

The growth in the volume of EU legislation has led to the emergence of a 'Europe hits home' literature—also known as 'Europeanization literature'— assessing national implementation of EU policies (see Mastenbroek 2005). This literature shows that EU policy has an independent effect on national policy outcomes, but that these effects differ from one setting to another (see discussions in Börzel and Risse 2006; Bulmer and Lequesne 2005; Mastenbroek 2005). Whereas this finding is uncontested, there are still many lacunae in our understanding of how EU policy and the European environment operate and change the national political context. As Börzel and Risse (2006:488) note: '[w]e hardly know anything about how the emergence of a European structure of political and societal interest representation impacts on the processes of political contestation and interest aggregation in the member states.'

Several scholars have argued that more theorization is needed in order to make sense of the interrelationship between European and national policy developments (e.g. Featherstone and Radaelli 2003; Olsen 2006). This book has taken up the challenge, opening a new venue of research: how entrepreneurship contributes to shaping the way that European developments play into national policy development. Because the Europeanization literature tends to see the relationship between EU policy and national policy change as a direct result of structural or institutional pressure, it fails to capture the importance of entrepreneurship. Entrepreneurship has attracted increasing attention in research on EU policymaking (particularly through the work of multi-level governance scholars), as well as in studies of national policymaking (network scholars in particular). However, there has not been a similar development with respect to how EU policy affects national policymaking.

The literature that underlines the structural importance of the EU has assumed, implicitly or explicitly, that the more coercive powers that are granted to the EU, the more important will EU steering be to the development of national policy (see Risse et al. 2001; Mastenbroek 2005). Our case studies have shown that that this is not necessarily the case. Even if the EU possesses considerable authority, its steering signals can be highly unclear and open to interpretation and negotiation. The case-study examples of 'Unpredictable EU governing' have made clear the significant room for interpretation, persuasion and negotiation in such situations.

Due to the competing authority structures involved, national-level actors may find themselves forced to perform entrepreneurship. Moreover, the entrepreneurship that emerges in such situations—Shrewd Lawyer entrepreneurship—will serve to enhance the entrepreneurs' impact on the national policy outcome. Because it is mostly civil servants who act as Shrewd Lawyer entrepreneurs, they gain power as a result of the significant coercive powers granted to the EU. Here the Norwegian CCS case is probably exceptional: politicians acted as entrepreneurs because the issue had such extraordinary high political importance. Normally, politicians will not have the capacity to take the lead in such lengthy negotiations. Thus, not only does EU steering represent a direct increase in the power of the EU over national policymaking: also the medium—the administrative staff—gains power thereby. However, this conclusion holds only for situations where the issue in focus is subject to considerable inter-EU disagreement and ambiguity. The EU may well have a greater direct impact when it is granted considerable formal authority *and* the environment is dominated by one policy model, as with the EU Emissions Trading System (see Boasson and Wettestad 2013).

Second, this book has added to the institutional perspectives on Europeanization. The 'goodness-of-fit' approach is explicit in its assumption that impulses for change will always stem from the EU. A certain degree of misfit is taken as a necessary, albeit not sufficient, condition for national change (Börzel and Risse 2003:58; see also Börzel and Risse 2006:491). However, we have seen that when the European environment is characterized by institutional conflict, there will be no direct relationship between goodness of fit and national policy outcomes. In such situations, national policy will not change unless highly skilled national entrepreneurs see it in their interest to bring the EU aspect into the national policy processes, and perform Importer entrepreneurship. Interestingly, entrepreneurs have enjoyed considerable leeway in interpreting the European situation: it is the entrepreneurs' view on Europe, and not actual European developments, that come to affect the development of national policy. Our case studies have provided examples of successful Importer entrepreneurship—but that is not to say that *all* opportunities will be exploited.

Moreover, it is not only misfit situations that can provide the backdrop for entrepreneurial activities: also situations of *fit* may spur entrepreneurship. Entrepreneurs who can depict their solution as being in line with EU developments, only even more forward-looking, may gain significant impact. Such 'Fashion Queen' entrepreneurship played a major role in the CCS case. That this was the sole instance of this kind of entrepreneurship in the four cases studies indicates that such entrepreneurship is less common than the three other types. Still, there is reason to expect that it can be exercised in various other issue-areas. For instance, Norwegian actors could have launched such arguments in relation to the energy requirements in the building code and the green certificate scheme. In both cases, Norway could have been said to be at the forefront of European development, but no actors chose to use this rhetorical technique in order to strengthen their policy impact.

The three entrepreneurial mechanisms with a European character (the Spider is national) create different ties between EU policy and national policymaking. Fashion Queens ensure that national policymaking is pitched as being developed in competition with other European leaders: hence, EU development becomes more important for national policy development than otherwise. Shrewd Lawyers can have considerable influence on both the strength and the content of the formal EU steering, although the entrepreneur will lose power after agreement is reached with the EU. In contrast, Importer entrepreneurs may interpret the situation without creating stronger connections between national-level and EU-level developments: this entrepreneurship is all about the interpretation of the European situation, and it is not necessarily closely aligned to the actual situation in the European environment.

Nonetheless, our case studies indicate that, with the exception of EU governing situations, the European environment seldom influences the development of national policy directly: it acts by altering the power sources of national actors. The active entrepreneurship involved makes it hard to pin down the actual importance of the EU. For instance, adaptation to the EU guidelines on state aid had the puzzling effect of strengthening the professional logics of the governmental organizations that ensured this, and not the professional logic that dominated the guidelines. Further, entrepreneurs were able to frame green certificates as the most profitable and innovative European approach to renewable state aid, whereas it was actually feed-in schemes—not green certificates—that diffused and dominated in Europe. Also with CCS, entrepreneurs framed European pollution regulation practices in a way that deviated strongly from the actual situation in EU member states. All these examples show that European impulses may be radically altered by national entrepreneurs.

8.5 ENTREPRENEURSHIP PARADOXES

This book has also inquired into the relative importance of social and entrepreneurial mechanisms when it comes to influencing policy development. Both political scientists and sociologists tend to acknowledge that the importance of entrepreneurship is limited—but few studies on entrepreneurship have actually explored the relative importance of entrepreneurship and other mechanisms (see Hardy and Maguire 2008:199, 207; Roberts and King 1991:172; Mintrom 1997:738; Bakir 2009).

Our case comparisons reveal a complex relationship between entrepreneurial and social mechanisms. Entrepreneurship played an important role in three out of the four cases (CCS, renewable electricity and buildings), significantly shaping policy development. In the three cases, social mechanisms in the two national fields (organizational and the political) incentivized entrepreneurial activities, and the European environment created entrepreneurial opportunities. Moreover, entrepreneurship ensured changes in the functioning of the main societal mechanisms. For instance, we have seen how political entrepreneurship contributed to the persistent Legislature Governing mechanism in the case of CCS, and corporate entrepreneurship ensured that the case of renewable electricity was heavily influenced by the Turf Battle mechanism. Better specification of the delicate inter-relationship between entrepreneurial and social mechanisms can help to modify the pluralist and network approaches of policy studies that see entrepreneurship as *the* key driver of national policy developments.

These conclusions are also relevant to the sociological neo-institutional literature, where it is commonly argued that entrepreneurship will be at its most significant in weakly institutionalized settings, in situations of institutional conflict or in crisis (see Battilana et al. 2009:74; Fligstein 2001a; Fligstein and Mara-Drita 1996; Hardy and Maguire 2008; Hoffmann [1997] 2001; Leblebici et al. 1991; Fligstein and McAdam 2012:51). These authors underscore how institutional uncertainty, confusion and the search for political or practical solutions provide entrepreneurs with opportunities. The arguments are sustained by various single case studies, including Leblebici et al.'s (1991) study of institutional change in the early years of the US broadcasting industry, and Fligstein and Mara-Drita's (1996) study of the creation of the European Single Market. However, contributors to this debate have themselves expressed self-criticism of their tendency to conduct research only on emerging fields (Battilana et al. 2009:74). As Fligstein and McAdam (2012:180) put it: 'the idea that the only moment in time that people have agency is when they are helping to form new social fields is odd to say the least'. Moreover, the neo-institutional scholars have suggested many situations in which entrepreneurship shall be more common, but the criteria they suggest are hard to apply empirically; many different situations may fit the descriptions of institutional conflict or crisis, uncertainty, confusion, or marked by a search for political or practical solutions. Our case studies have shown that the entrepreneurship that unfolded in a situation of stark conflict (CCS) had a radically different character than in a situation with many co-existing professional logics (buildings) or in a situation of uncertainty and a search for policy solutions (the case of renewables).

Rather than trying to develop a new general description of situations that foster entrepreneurial activities, the multi-field framework specifies how the structural and institutional features in two specific national fields (the organizational and the political) influence the volume of entrepreneurship. We have seen that conditions of the organizational field are generally the most important when it comes to the volume of entrepreneurship, simply because politicians exercise entrepreneurship less frequently than do organizational field actors. Sociologists Thomas Lawrence and Roy Suddaby (2006:247) have argued that the creation, maintenance and disruption of social practices within an organizational field will always require substantial entrepreneurial efforts. This book sees entrepreneurship as fairly common phenomenon in policy development, but not as a requirement for either policy emergence or change.

The multi-field framework offers a systematic understanding of how political and organizational fields separately and in conjunction create entrepreneurial opportunities, influencing the chances of shaping policy development. Neither neo-institutional sociologists nor policy scholars interested in entrepreneurship have explored this before. Our case studies indicate that politicians have only limited capacity for entrepreneurship; hence very few political issues will gain entrepreneurial attention on the part of politicians. This finding may lead us to say that the bulk of scholars that have explored entrepreneurship made a good choice in largely ignoring political entrepreneurship, and focusing more on entrepreneurship performed by business organizations (Greenwood and Suddaby 2006), corporations (Lounsbury and Crumley 2007; Mazey and Richardson 2006), environmental organizations and other social movements (Davis et al. 2005; Lounsbury 2007; Sabatier and Jenkins-Smith 1993a). Importantly, however, our case studies clearly show that political entrepreneurship can significantly enhance the influence of the political field on a given policy area. Hence, even though it is rare, that does not mean that it is without significance.

Our case studies also shed new light on the relationship between the volume of entrepreneurial activity and its effectiveness. Indeed, entrepreneurship seems to be least efficient when it is exercised the most, and the most efficient when exercised the least. In the CCS case, the entrepreneurs contributed to bring about a Norwegian CCS policy in the first place, but both the actual character of the policy and the instability of the policy area were more the consequence of social mechanisms than of deliberate design on behalf of the entrepreneurs. With renewable electricity, entrepreneurship was not important for the emergence of a policy in the first place, but the shift towards a market measure resulted from entrepreneurship. The emergence of an energy policy for buildings hinged on several instances of entrepreneurship, which, despite its modest and *ad hoc* nature, was of paramount importance for the policy development. Even though entrepreneurs invested tremendous efforts in CCS, they succeeded only partly. Entrepreneurship was important to the outcome in the case of renewable electricity, but it took a very long time before success was achieved. Concerning an energy policy for buildings, entrepreneurs made some achievements, but if they had engaged more, the results would probably have been far more significant.

Why should issues where entrepreneurial success is hardest to achieve tend to attract the most entrepreneurial activity? It seems that, when faced with strong resistance, entrepreneurs are forced to engage strongly in order to achieve their objectives; and the more they have invested the more they will commit to continuing their entrepreneurial activities. Entrepreneurship patterns can be path-dependent: as long as political competition is prevalent, politicians are not free to delve into new issues. They must first resolve the political hurdles that previous politicians made into important symbolic issues.

When there is considerable leeway for entrepreneurial action, the issue will seldom attract much initial attention from entrepreneurs—but if entrepreneurs are mobilized, their seemingly swift success will motivate them to move on to new tasks, rather than ensure that their efforts actually lead to the intended results. Hence, the large pool of entrepreneurial opportunities in issue-areas with fewer constraints on policy development is rarely exploited to the full.

Moreover, whereas the literature on entrepreneurship tends to focus on how entrepreneurs enable or spur change, our case studies show how entrepreneurship may also make processes of change more cumbersome. This was particularly clear in the case of renewable electricity, where entrepreneurship on the part of the utilities and environmental groups led to lengthy political processes, while all actual investments remained halted for a decade. Lack of entrepreneurial activity hindered dramatic shifts in the renewable heating policy, but also sustained actual changes in industry practice: eventually, the scheme changed in a way that meant enhanced profitability for the industry actors. Hence, actors who aim to produce ambitious climate policies are faced with a paradoxical challenge: if they choose to act as entrepreneurs and promote their preferred policy solution, that may prolong the political processes significantly, creating instability that in turn hinders a low-carbon transition.

These findings clash with common assumptions in the political science literature on entrepreneurship. Although this is seldom stated explicitly, political scientists often assume that the more entrepreneurial activities that are exercised, the more likely is it that the entrepreneur will gain power to shape the policy outcome. Among the writers who assume that the most active entrepreneurs will be the most powerful we find Robert Dahl, who argues that, although a gifted entrepreneur might not exist in every political system, where he appears he will make himself felt (Dahl 1961:6). Some network theorists, among them Sabatier and Jenkins-Smith (1993b), have indicated that actors who mobilize networks to defend their 'core beliefs' will be powerful. However, the effectiveness of entrepreneurship seems to be inversely related to the energy devoted: paradoxically, entrepreneurship is least efficient when it is exercised the most.

This book has shown that, while we need to take entrepreneurship into account in order to understand process of national climate policy, it is essential to realize that entrepreneurship is not merely something that results from the initiatives and skills of outstanding actors. Different situations create different kinds and numbers of entrepreneurial opportunities. Moreover, intuitional and structural conditions in multiple fields will influence how and to what extent entrepreneurship may influence policy development.

8.6 NEW AGENDA FOR CLIMATE-POLICY RESEARCH

Chapter 1 presented the state of the art as regards climate-policy studies, clearly showing that this political science literature is still in its infancy. The UN International Panel on Climate Change (IPCC), Working Group III report on National and Sub-national Policies documents steep growth in national climate policies in the 2000s, but also that we lack insights into the factors that drive policy emergence and change. Moreover, it is evident that we need to specify and conceptualize more elaborate theory frameworks that can capture the true-life complexities of climate policy developments.

The multi-field approach opens up a new research agenda for studies of climate policy, based on a complex view on causal relationship. The case studies presented here have shown that explanatory factors are interdependent and that the importance of certain mechanisms may vary over time or place. Moreover, few causes are absolute: most causes are probabilistic (Mahoney 2003:344; Hall 2003:281).

Policy researchers have traditionally operated with rather short time-spans, but today is it increasingly recommended that policy assessment be conducted over a period of at least 10 years (see Pierson 2004:114; Sabatier and Jenkins-Smith 1993b). Our case studies confirm that the causal processes that shape climate-policy outcomes tend to play out over fairly lengthy time-spans. We see many examples of path dependency, where different factors are at work over time, often in slow-moving processes. In such situations, explanations need to focus on sequences of events, some of which may foreclose certain paths in the development and steer the outcome in other directions (George and Bennett 2005:212; Pierson 2004; Streeck and

Thelen 2005). Future studies of climate policy are advised to regard national policy outcomes as a function of events that unfold over time.

Further, the multi-field framework can enable climate-policy researchers to discover a variety of possible different causal patterns. Rather than searching for simplicity and elegance in causal explanations, climate-policy research should be based on the insight that developments in climate policy are characterized by *equifinality:* multiple causal paths may lead to the same outcome (Ragin 1987; Mahoney and Goertz 2006:11). Our case studies show that a similar policy outcome can result from different mechanisms. For instance, there were radical differences in the causal factors that led to one market instrument—the green certificate scheme—and another market instrument—energy certification of buildings.

Moreover, climate policy researchers need to be sensitive to complex interdependencies, between the explanatory factors or between the explanatory factors and the outcomes (Hall 2003:383; Rueschemeyer 2003:315). This book has detected and specified interdependencies between entrepreneurial and social mechanisms within the field, and interdependencies between fields. Context matters: the impact of one mechanism is rarely independent of other mechanisms (Hall 2003). This is particularly true of policy development over fairly lengthy stretches of time: for instance, we have seen that European discussion about a green certificate scheme in 2000 failed to influence many other countries than Sweden and Norway, and the effects in Norway did not materialize until almost 10 years later, due to a combination of entrepreneurship and social mechanisms.

Alexander L. George and Andrew Bennett (2005:111; see also Bennett 2008:711) have convincingly argued that in order to produce valid explanations we need to cast the net of alternative explanations widely. The multi-field approach is based on an extensive synthesizing of existing theory approaches (see Collier et al. 2008:160). It is a theoretical pluralist framework in the sense that it applies and combines different theories, taking seriously the basic fact that all observation is theory-laden. Most social science theories have been overly focused on one field only—national policy theories have tended to focus on national factors, EU theories on EU-internal factors, and so forth. The case studies presented here have shown how the relative importance of the two national fields and the European environment varies across issue-area and time.

Theoretical pluralism has helped us to understand how mechanisms at work in different fields may affect each other, enabling us to grasp better the complexity of real-life causal processes (Patomäki and Wight 2000:227; Danermark et al. 2002:63). It is my hope that the theory pluralism of this book will help to foster communication across different scholarly communities—which in turn is a precondition for accumulating deeper knowledge on the development of climate policy. Further, climate-policy studies will profit from becoming more engaged in general theory discussions in policy studies, and the social sciences at large.

8.7 FUTURE RESEARCH

The framework presented in this book is not a readymade product. It represents a cautious start towards enhancing the analytical quality of political science research on national climate policy. The multi-field framework may serve to underpin the development of a more coherent, comparative research programme for national climate policy, fostering more systematic comparisons across issue-areas as well as countries. Future application of the framework to new issue-areas and new countries will contribute to refining and improving it.

The multi-field framework draws on many theory traditions, but the comparative studies presented in this book come from one country only. Moreover, it is assumed that the framework can more readily be applied in its totality in other EU and EEA member states than in countries elsewhere. Hence, priority should go to comparative studies on more European countries. As the multi-field framework requires resource-demanding, historical studies it will be best to begin with studies of a few countries, expanding the number later. If we stick to the issue-areas explored in this book, we can create bigger samples for cross-country comparisons (on CCS, renewable energy and buildings). However, we also need to enlarge the number of climate-policy areas: particularly important areas to explore are carbon regulations (emissions trading, CO_2-taxes or pollution regulation) and transport policies (public transport, electric vehicles, urban planning, etc.). It would also be of considerable interest to apply the multi-field framework to climate policy adaptation issues, and to compare similarities and differences across the adaptation–mitigation divide in climate policy.

It should be a long-term objective to develop multi-field assessments not only of single climate-policy areas but of the broader climate-policy portfolios of various countries as well. This can enable more holistic cross-country comparisons. Further, we need studies on all regions of the world in order to gain a better understanding of which parts of the framework have universal applicability and which will need to be adjusted to non-European contexts.

Perhaps history will show that some of the multi-field framework components were inadequate, lame or directly wrong. That said, there is reason to be satisfied if introduction of the framework can serve to underpin fruitful debates on comparative climate policy theories and methods.

9 Advice to Policymakers and Stakeholders

9.1 INTRODUCTION

Actors with a good understanding of climate policymaking have a better starting point for influencing climate policy. This book speaks to scientific communities of political science and sociology scholars, but it also aims to make policymakers and stakeholders more able to develop sound strategies for climate policy. It is not intended as a cookbook on how to develop ambitious national climate policy: it offers none of the 'silver bullet', 'quick-fix' or 'one-size-fits-all solutions' that so many have called for in the climate-change debate. Rather, it acknowledges the complexities of contemporary national policymaking, and then offers some suggestions for practitioners.

Anthony Giddens (2009:2) argues that, because the dangers posed by global warming are not tangible, immediate or visible in the course of everyday life, most actors will sit on their hands and do nothing of a concrete nature about them—a phenomenon he rather immodestly calls the 'Giddens paradox'. However, in this book we have seen that there are many policymakers, industry representatives and other stakeholders that that are *not* paralysed by the Giddens paradox. National climate policymaking may at times be smooth sailing—whereas under other circumstances even the best-intentioned efforts may fail to produce clear and stable policy outcomes. This can be frustrating for all actors with a stake in the policy development processes—not least because it seems as if the issues that attract the most entrepreneurship are those where it may be hardest to succeed in inducing policy change.

This chapter will summarize the overall insights of this book into the following seven pieces of advice to policymakers and stakeholders who seek to influence and change national climate policies:

1. Map the organizational field(s)!
2. Map the position of the issue in the political field!
3. Work with the instructional logics, not against them!
4. Get the priorities right!

5. Be aware of the advantages and pitfalls of 'garbage can' processes and politicization!
6. Be aware of entrepreneurship opportunities relating to Europeanization!
7. Always look on the bright side!

The seven points of advice will be discussed in relation to Anthony Giddens' *The Politics of Climate Change* (2009) and Hugh Compston and Ian Bailey's *Climate Clever: How Governments Can Tackle Climate Change (and Still Win Elections)* (2012). Both books present advice based on the high social-science expertise of the authors. Neither book offers analytical frameworks or systematic empirical studies.

9.2 MAP THE ORGANIZATIONAL FIELD(S)

Compston and Bailey (2012) recommend putting emphasis on the climate issues that are least likely to meet protests from industry and political majorities. This requires mapping of the actors with something at stake in an issue. In order to influence a process of climate-policy development, it is useful to know who the other actors are and what interests they have flagged previously. But this is not sufficient: it is also essential to understand how they think, how committed they are, and what powers they may muster.

The first step is to identify the organizational field in which an issue is embedded and explore the institutional and structural character of this field. In these efforts, we ought to follow Giddens (2009:120) advice of not 'demonizing the industry lobbies', but rather understand the variance in industry positions. Nor should we overlook the potential powers of governmental administrative organizations and their staff. Further, it is important to question whether the field is dominated by any of the three climate-policy logics: minimization of societal costs; market measures; or technology development? Or do all logics exist in conjunction, creating conflicts and inconsistencies among the views of the fields' actors? Knowledge of the institutional character of the field will help in understanding how key field actors think, how they understand their role, what they perceive as rational action, and what really is at stake for them. It will also tell a lot about how they may react to a given policy proposal and how that proposal should be framed in order to reduce the level of resistance.

Second, which actors control the structural resources of the field, the formal decision-making authority and important technical information? Are such resources concentrated within one or two organizations, or are they widely distributed among a whole range of actors? Assessment of the structural features will help to make clear the magnitude of resistance or support to be expected, whether there will be significant support from field-level actors, or if it will be necessary to approach the political field in order to gain support. In fragmented fields, it will probably always be possible to find someone willing to support a given cause, but no actors will be in a position to control considerable numbers of other field actors.

Third, many climate-policy issues have a prime rooting in one organizational field (like those assessed in this book), but some are influenced by many different fields. For instance, emissions trading systems and other carbon regulation tools will tend to be rooted in the organizational fields of stationary energy and energy-intensive industries, as well as petroleum. The more organizational fields that are engaged in a policy issue, the more complex and opaque will the policy development processes tend to be.

9.3 MAP THE POSITION OF THE ISSUE IN THE POLITICAL FIELD

Giddens as well as Compston and Bailey tend to present climate-policy discussions as a conflict between 'the willing' and 'the unwilling'. For instance, Compston and Bailey (2012:85) underline the importance of getting more 'climate-friendly' politicians elected. In fact, viewing politicians as being *either* negative *or* positive forces in climate-policy development is somewhat misleading. This book has shown how the chances of receiving political acceptance will vary from one issue-area to another, often due to differences in the structural and institutional character of the political field. It is important to understand which politicians have a stake in the issue and what position they may have taken previously—but such knowledge is not sufficient for assessing what could or should be done in order to influence actual decision-making: it is essential to understand how the structural resources are distributed and whether the issue in question is entangled in political competition or more random 'garbage can' processes.

As underlined by Compston and Bailey (2012) some policy solutions will need a parliamentary decision in order to be adopted, whereas others may be decided by individual ministers, or by a vote among all the ministries in the government. If a specific ministry has formal authority to take decisions in relation to an issue, it may be of little use in persuading parliamentary committees to support a specific viewpoint: it would probably be better to target the political leadership in the ministry. In some instances, the distribution of decision-making authority is rather fixed, but our case studies have also shown that this may change over time. If an issue becomes politicized, it may also be lifted out of the domain of the ministry and into the parliament. This may lead to radical change in the distribution of power in the political field, particularly if there is a minority government. Also important is the parliamentary situation: if the government has a majority, parliamentary power is less likely to change than under a minority government.

The ruling political logic has major implications for how much attention and energy politicians will be willing to devote to a given proposal from a stakeholder. If the proposal is related to on-going political competition, it may soar to the top of the agenda. This may be an advantage if it plays well with the competition issue (renewable energy and energy security is often a good match), whereas it may be negative if they contrast (climate mitigation and soaring energy prices tends to be a bad match). But because very few issues are competed over, the political engagement will probably depend on timing (whether the political agenda is hectic at the moment) and the personal interest of the politicians one approaches. Only if the idea relates to a competitive issue are politicians likely to be willing to devote much time to it, and only under such circumstances can they be expected to pay attention to the details of the proposal. Otherwise, the most one can hope for is to gain some symbolic support. However, symbolic decision can be crucial, particularly if the civil servants are inclined to act upon weak political signals.

9.4 WORK WITH THE INSTRUCTIONAL LOGICS, NOT AGAINST THEM

Many environmental groups, but also business groups, are primarily interested in the actual results of their policy proposals, for instance in terms of carbon mitigation or increased investments in low-carbon technology and practices. Yet, climate-policy discussions often get lost in endless discussions about how things should be done. Neither Giddens nor Bailey and Compston pay attention to this phenomenon. However, the case studies have shown how disagreement concerning policy design can have severe ramification for policy development and implementation. Conflicts between professional, institutional logics often play a key role. Hence, it is wise to try to avoid such conflicts wherever possible, and to develop strategies for dealing with them if needed.

This book has shown that there is no consensus on what good climate policy is. A policy idea may be perceived as ingenious in the light of one institutional logic, and insane in the light of another. For instance, from the technology development perspective of an engineer it will be a very good idea to develop technology-specific renewable support schemes that give investors long-term stability for refining and improving a broad range of renewable energy technologies (like feed-in schemes). Business economists or other actors embedded in a market logic may find this downright stupid: as they see it, what is needed is a policy that can create competition between projects and actors so as to ensure that the best projects are developed and the actors willing to take the largest economic risks are rewarded (like green certificates).

In some instances, it may seem as if the discussion is about carbon mitigation but the actual centre of the deliberation is something else, with each actor trying to defeat each other's professional logics and persuade the other to shift positions. Norwegian climate policies have been rife with such conflicts between the market logic and the minimizing societal costs logic; and EU discussions have featured such conflicts between market logics and technology development logics. These discussions are perfectly legitimate, and indeed, actors may eventually change positions and come to understand

each other better. Unfortunately, however, there are generally few enduring consensus solutions available, so actors who initially aimed at ensuring
GHG mitigation may get lost in disagreement over measures: the result is mere
talk and little action. Hence, actors seeking to engage in policy development
are advised to pay attention to how their proposals fit with dominant logics,
and to try to avoid evoking ideological conflicts as much as possible. After
all, it will often take time to change the professional logics that dominate
fields, and the institutional patterns of fields are a result of many factors, not
only climate-policy processes.

9.5 GET THE PRIORITIES RIGHT

Giddens, Compston and Bailey pay scant attention to the importance of setting priorities. There remains much to be done in order to mitigate climate
change, and it can be hard to distinguish the important issues from the
less important ones. It may also very well be that the issues which climate-
policy enthusiasts find the most important will be those that are the hardest
to change: for instance, many industrialized countries have powerful industries with vested interests in carbon intensive activities and high pollution
records. When such industries, as we have seen with respect to the Norwegian petroleum industry, have the upper hand in a national organizational
field, they will have considerable power to resist political campaigns.

It can be argued that it is of fundamental importance to nonetheless promote certain unpopular policy solutions targeting such industries. This may,
as we have seen in the Norwegian case, ensure that at least one climate issue
is high on the political agenda. Whereas this strategy is not likely to immediately lead to positive results in the issue at stake, it may create substantial
political commotion and competition—even a huge space where action can
actually happen: we have seen that related areas of climate policy may stand
to profit from problems in high-profile areas, by being granted symbolic support in governmental declarations and overarching parliamentary decisions
on climate policy.

However, such a strategy may also backfire: it may require so much entrepreneurial energy that people will not have resources left to deal with all
other parts of the grand climate-policy agenda. It is essential to try to think
ahead about the possible long-term consequences of various climate-policy
issues, and how these policy areas may influence each other over time.

9.6 BE AWARE OF THE ADVANTAGES AND PITFALLS
OF 'GARBAGE CAN' AND POLITICIZATION!

Giddens, as well as Compston and Bailey, recommends cross-party alli-
ances on broad climate deals. This may be good advice, but it overlooks
the fact that, irrespective of potential package deals, different issue-areas

tend to be dominated by different political logics—and this will have profound influence on how politicians deal with different areas of climate policy.

Garbage Can logics create only shallow and ephemeral political commitment, and it is not likely that the broad range of politicians will actually understand and commit to the issues. This may seem like a very bad thing—but it also comes with some advantages. Firstly, individual politicians in ministries with authoritative decision-making rights may play an important role in such issue-areas. They need not go out of their way in order to exert influence: it is enough that they simply do their job and make use of the privileges that come with their positions. If decision-making power in the issue-areas is distributed, it is possible to shop between decision-making arenas and opt for the one where success is most likely.

Note however, that entrepreneurs risk to be fooled by their swift success in Garbage Can situations: even though they may readily get support for their ideas, this does not necessarily mean that the political party or politician that they have convinced will act in a similar way the next time around. Rather, it is likely that another issue and other concerns will be paid attention to at the next cross-road. Hence, stakeholders are advised to be sensitive to whether their input has been treated in a garbage can fashion: is this is the case they will have to mobilize in the same way at every decision opportunity.

Giddens (2009:12) recommends avoiding 'making political capital out of global warming'. In saying this, he ignores the fact that politicization of a climate-policy issue may be precisely what is needed in order to get climate policy on the agenda in the first place. He also argues that global warming needs to be 'a front-of-the-mind-issue', at 'the core of the political agenda' (2009:71). It is hard to envisage how the latter could be possible unless politicians aim to make political capital out of the issue. Indeed, politicization will often be the only way to get politicians to develop long-lasting commitments to climate mitigation.

There are, however, also pitfalls to politicization. Once an issue gets politicized, the political development will be shaped by the political competition, not a coherent, professional approach to problem solving. It may take time to politicize an issue and the consequences are uncertain. All actors should, in advance, think through whether it is necessary to spur political competition in order to get it their way with the politicians. Actors that conclude that the shallow commitment available through 'garbage can' will probably not do the trick, best prepare for lengthy entrepreneurial engagement.

Political competition means that actors outside the political realm will have little ability to influence the policy processes. The main focus will be set on winning political battles; this may imply adoptions of compromises that are less than perfect when it comes to their ability to achieve carbon mitigation.

9.7 BE AWARE OF ENTREPRENEURSHIP OPPORTUNITIES RELATING TO EUROPEANIZATION

All actors with a stake in climate policy ought to check out their chances of using European features to promote their cause. Perhaps more importantly: actors ought to try to understand how others use, or may use, European arguments to increase their clout. Even though the EU has developed a substantive climate policy, EU authority over national climate policymaking is limited. Still, it can be a very forceful argument that your view is in line with EU requirements. Moreover, there are at all times multiple negotiations underway within EU organizations on whether current national practices are in compliance with EU practices.

Not all actors have the opportunity to follow EU developments closely and check whether actors claiming that they promote EU policy actually do so, or whether they have been 'creative' in their interpretation. But everyone can adopt critical inclinations to such arguments, checking the basics when it comes to the strength of EU rules in the area.

9.8 ALWAYS LOOK ON THE BRIGHT SIDE!

Even though mapping of the multi-field landscape can give better foresight, it is no panacea: actors may still act in mysterious ways and do unexpected things. Compston and Bailey (2012:112) are certainly right in arguing that '[p]olitics is a messy and unpredictable business'. This book has highlighted the complexities of the national climate-policy development processes—perhaps not what actors with good ideas and hopes for swift social transformation towards a low-carbon society want to hear. Highlighting only the complex and cumbersome nature of climate policy may create apathy and despair, which will not bring us closer to coping with climate change.

Giddens, as well as Compston and Bailey, do a good job of arguing that a more positive framing of climate-policy solutions is needed: we should not act out of fear alone, we need to believe in the benefits of climate policies. I hold that it is just as important that climate-policy activists and policymakers believe that there is actual chance of getting ambitious climate policies adopted. I will encourage readers to focus on the bright sides of requiring a deeper understanding of climate-policy development. Most important of all: the multi-field approach has shown that there will usually be ample room for creativity and entrepreneurial engagement. The better actors know the climate-policy landscape, the more likely are they to spend their resources well.

Interviewees

The following 95 persons were interviewed in connection with the PhD work that underlies this book. Most interviews were conducted with one person at a time, but a few were interviewed in groups of two or three.

Ådland, Hans Magne, Energibedriftenes Landsforening (later: Energy Norge/Norwegain Energy), Oslo, 8 October 2008

Aas, Agnar, Norwegian Water Resources and Energy Directorate, Oslo, 18 June 2009

Andersen, Birgitte, EFTA secretariat, Brussels, 25 February 2008

Andersson, Bosse, Vattenfall, Stockholm, 13 March 2009

Arnstad, Eli, Director Enova from 2001 to 2007, Oslo, 11 December 2008

Asheim, Kari, Norsk Bioenergiforening, Oslo, 30 July 2007

Bergesen, Birger, Norwegian Water Resources and Energy Directorate, Oslo, 13 September 2006

Bergflødt, Lise, Skanska, Oslo, 7 November 2008

Bernsen, Johanna, European Commission, DG Competition, Brussels, 26 February 2008

Blanken, Joris van den, Greenpeace, Brussels, 22 June 2009

te Bos, Jan, EURIMA, Brussels, 25 June 2009

Bowie, Randall, Rockwool from September 2007; previously European Commission, DG Energy; Brussels, 26 February 2008

Bratland, Sjur, Hydro, Oslo, 3 May 2007

Broli, Erlend, Statkraft, Oslo, 8 October 2008

Bysveen, Steinar, Energibedriftenes Landsforening, later renamed Energi Norge (Norwegian Electricity Industry Association/Energy Norway), Oslo, 17 June 2009

Christensen, Dag, Hydro, Oslo, 3 May 2007

Clayon, Marianne, ETA Surveillance Authority (ESA), Brussels, 29 February 2008

Dagestad, Brita, National Office of Building Technology and Administration (the Building Agency), Oslo, 19 February and 20 August 2006

Dahl, Agnethe, Ministry of the Environment, Oslo, 15 June 2009

Dokka, Tor Helge, Sintef Building and Infrastructure, Oslo, 5 October 2008

Engebretsen, Marit, Norwegian Ministry of Petroleum and Energy, Brussels Delegation, Brussels, 26 February 2008, 23 June 2009

Enoksen, Odd Roger, Minister of Petroleum and Energy (17 October 2005–21 September 2007), Oslo, 2 February 2008.

Eriksen, Henrik, Norwegian Ministry of the Environment, Brussels Delegation, Brussels, 23 June 2009

Ettestøl, Ingunn, Enova, Oslo, 5 June 2007

Faraday, Frank, European Construction Industry Federation (FIEC), Brussels, 28 February 2008

Foquet, Dörte, European Renewable Energies Federation (EREF), Brussels, 28 February 2008

Frisvold, Paal, Bellona, Oslo, 23 July 2009

Fuglseth, Geir, Naturkraft, Lysaker, Norway, 25 May 2009

Gjerset, Marius, ZERO, Oslo, 3 June 2009

Gjerstad, Frode Olav, Enova, Lysaker, Norway, 3 October 2006

Goodall, John, European Construction Industry Federation (FIEC), Brussels, 28 February 2008

Graff, Oscar, Aker Clean Carbon, Lysaker, Norway, 20 May 2009

Grini, Gunnar, National Office of Building Technology and Administration (the Building Agency), Oslo, 19 February and 20 August 2006

Gundersen, Mari Hegg, Norwegian Water Resources and Energy Directorate, Oslo, 6 June 2007

Håbrekke, Øyvind, Political Advisor/Deputy Minister, Ministry of Petroleum and Energy (19 October 2001–27 January 2004); Deputy Minister, Ministry of the Environment (18 June 2004–17 October 2005), Oslo, 29 June 2009

Haga, Åslaug, Minister of Local Government and Regional Development (17 October 2005–21 September 2007); Minister of Petroleum and Energy (21 September 2007–20 May 2009), Oslo, 20 May 2009

Hammer, Erik, Grønn Byggallianse (Green Building Alliance), Oslo, 19 November 2008

Håndlykken, Einar, ZERO, Oslo, 31 January 2008

Haugan, Bjørn-Erik, Gassnova, Oslo, 30 April 2009

Hedenstrøm, Claes, Vattenfall, Stockholm, 13 March 2009

Hedstrøm, Jenny, Energimyndigheten (Swedish Energy Agency), Stockholm, 11 March 2009

Hercsuth, Andrea, European Commission, DG Transport and Energy, Brussels, 27 February 2008

Holm, Marius, Bellona, Oslo, 10 June 2009

Isachsen, Olav K., Norwegian Water Resources and Energy Directorate, Oslo, 13 September 2006

Jæger, Per, Boligprodusentene (Association of Dwelling Producers), 20 October 2008

Karlsen, Tom, Ministry of Petroleum and Energy, Oslo, 24 October 2008

Juhler, Heidi, Norsk Fjernvarmeforening (Norwegian Association for District Heating), Oslo, 12 September 2007

Kismul, Ane Hansdatter, Norwea (Norwegian Wind Association), Oslo, 11 May 2007

Klimmann, Annette, ETA Surveillance Authority (ESA), Brussels, 29 February 2008

Konglevold, Synnøve, Member of the Storting 1997–2001, Vækerø (Norway), 11 November 2007

Koskimaki, Pirjo-Liisa, European Commission, DG Transport and Energy/DG Energy, Brussels, 25 June 2009 and 25 January 2011

Kroepelien, Knut, Norwegian Ministry of the Environment, Brussels Delegation, Brussels, 23 June 2009

Kumar, Sanjeev, WWF, Brussels, 22 June 2009

Legård, Jørgen, Byggenæringens Landsforening (Federation of Building Constructors), Oslo, 24 October 2008

Leistad, Øyvind, Enova (former Ministry of Petroleum and Energy), Oslo, 15 June 2007

Lier-Hansen, Stein, Norsk Industri (Federation of Norwegian Industries) (also Deputy Minister Ministry of Environment 2000–2001), Oslo, 15 June 2009

Lipponen Juho Eurelectric, Brussels, 25 February 2008 and 24 June 2009

Meyer, Raphael, ETA Surveillance Authority (ESA), Brussels, 22 June 2009

Myhre, Lars, Boligprodusentene (Association of Dwelling Producers), Oslo, 20 October 2008

Møller, Ulf, Statnett (Norwegian Transmission System Operator), Oslo, 17 January 2008

Nordlund, Per, Statkraft Sweden, Stockholm, 11 March 2009

Nygård, Per, Ministry of Local Government and Regional Development, Oslo, 12 February 2009

Olsen, Arne, Norwegian Water Resources and Energy Directorate, Oslo, 25 May 2007

Piel, Elo, Euroheat and Power, Brussels, 29 March 2008

Radmann, Trine, NHO Europe (Confederation of Norwegian Enterprises, Europe), Brussels, 27 March 2008

Rebo, Hans Petter, StatoilHydro, Oslo, 19 June 2009

Rødsjø, Are, Norwegian State Housing Bank, Oslo, 3 December 2006

Rosenqvist, Per, Statkraft Sweden, Stockholm, 11 March 2009

Rushe, Tim Maxian, European Commission, DG Transport and Energy, Brussels, 27 February 2008

Rusten, Birgit, Norske arkitekters landsforbund, Ecobox (Association of Norwegian Architects), Oslo, 20 December 2008

Sanderud, Per, ETA Surveillance Authority (ESA), Oslo, 19 June 2009

Schaefer, Oliver, European Renewable Council (EREC), Brussels, 25 June 2009

Sharpe, Dale, UK Brussels Delegation, Brussels, 23 June 2009

Skogseid, Inger Margrete, Ministry of Local Government and Regional Development, Oslo, 19 August 2006

Solheim, Marit, Ministry of the Environment, Oslo, 15 June 2009

Sørensen, Heidi, Member of the Storting, 2001–2007, Oslo, 23 August 2007

Steen, Hans van, European Commission, DG Transport and Energy, Brussels, 23 June 2009

Steensnæs, Einar, Minister of Petroleum and Energy (19 October 2001–8 June 2004), Oslo, 22 June 2007

Stokknes, Stein, Norske arkitekters landsforbund, Ecobox (Association of Norwegian Architects), Oslo, 20 December 2008

Størvold, Guri, Deputy Minister, Ministry of Local Government and Regional Development (17 October 2005–21 September 2007), Oslo, 19 August 2006

Strandskog, Tore, Norsk Teknologi (Norwegian Technology), 23 October 2008

Stubholdt, Liv Monica, Deputy Minister, Ministry of Petroleum and Energy (20 June 2008–27 March 2009), Oslo, 26 May 2009

Ticau, Silvia Adriana, European Parliament, Brussels, 24 June 2009

Tranholm-Schwarz, Bente, European Commission, DG Competition, Brussels, 26 February 2008

Tveitereid, Sigurd, Ministry of Petroleum and Energy, Oslo, 27 July 2007

Ulseth, Oluf, Statkraft (Deputy Minister, Ministry of Petroleum and Energy (18 June 2004–17 October 2005), Lysaker, Norway, 8 October 2008

Vernmark, Bengt, Statkraft Sweden, Stockholm, 11 March 2009

Vetlesen, Johan, Ministry of Petroleum and Energy, Oslo, 16 December 2008

Veum, Karina, European Commission, DG Transport and Energy, telephone interview, 28 March 2008

Vis, Peter, European Commission, DG Transport and Energy/DG Climate, Brussels, 22 June 2009

Vollsæter, Geir, Shell, Alston & Bird (previously Shell Norway and Shell International), Oslo, 10 October 2009

Warren, Andrew, Eurace, Brussels, 25 June 2009

Westgård, Geir, Statoil Brussels office, Brussels, 22 June 2009

Westrin, Henriette, Deputy Minister, Ministry of the Environment 2005–2007; Deputy Minister, Ministry of Finance, 2007–2009, Oslo, 18 June 2009

Zander, Joachim, ETA Surveillance Authority (ESA), Brussels, 29 February 2008

References

Aall, C., Groven, K. and Lindseth, G. (2007) 'The scope of action for local climate policy: the case of Norway', *Global Environmental Politics* 7(2):83–101.

Abbott, A. (2007) 'Mechanisms and relations', *Sociologica*, 2(1):1–22.

Acharya, A. (2004) 'How ideas spread: Whose norms matter? Norm localization and institutional change in Asian regionalism', *International Organization*, 58(2): 239–75.

Agder Energi (2007) *Årsrapport 2006*. Kristiansand, Norway.

Aker Clean Carbon (2010) Aker Clean Carbon. Available at: www.akercleancarbon. com/ section.cfm?path=417 (Accessed 20 January 2011).

Akershus Energi (2008) *Årsrapport 2007*. Rånåsfoss, Norway.

Allison, G. (2006) 'Emergence of schools of public policy: reflections by a founding dean', pp. 58–79 in R.E. Goodin, M. Rein and M. Moran (eds), *The Oxford Handbook of Public Policy*. Oxford: Oxford University Press.

Allison, G. and Zelikow, P. (1999) *Essence of Decision: Explaining the Cuban Missile Crisis*, 2nd edn. Reading, MA: Addison-Wesley.

Antill, N. and Arnott, R. (2004) 'Creating value in the oil industry', *Journal of Applied Corporate Finance* 16(1):18–31.

Bachke, N. (2003) 'Fra en generell til en selektiv boligpolitikk?' Thesis in political science, University of Oslo.

Bachrach, P. and Baratz, M.S. (1962) 'Two faces of power', *American Political Science Review*, 56:947–52. Reprinted in M. Haugaard (2002) *Power: A Reader*, pp. 28–37. Manchester: Manchester University Press.

Backe, I. and Flinders, M. (eds) (2004) *Multi-level Governance*. Oxford: Oxford University Press.

Bakir, C. (2009) 'Policy entrepreneurship and institutional change: multilevel governance of central banking reform', *Governance: An International Journal of Policy, Administration, and Institutions*, 22(4):571–98.

Bang, G. (2010) 'Energy security and climate change concerns: NACPOCs for energy policy change in the United States?' *Energy Policy*, 38(4):1645–53.

Battilana, J., Leca, B. and Boxenbaum, E. (2009) 'How actors change institutions: towards a theory of institutional entrepreneurship', *Academy of Management Annals*, 3(1):65–107

Baumgartner, F.R. and Jones, B.D. ([1993] 2009) *Agendas and Instability in American Politics*. 2nd edn. Chicago, IL: University of Chicago Press.

Bennett, A. (2008) 'Process tracing: a Bayesian perspective', in J.M. Box-Steffensmeier, H.E. Brady and D. Collier (eds), *Oxford Handbook of Political Methodology*. Oxford: Oxford University Press.

Berge, U.G. (2005) 'Petroleumsaktivitet i Barentshavet: Konflikt eller sameksistens?' Masters thesis, University of Oslo. Available at: www.duo.uio.no/sok/work. html?WORKID=32067 (Accessed 20 January 2011).

Bhaskar R. (1998) 'General Introduction', pp. ix–xxiv in M. Archer, R. Bhaskar, A. Collier, T. Lawson and A. Norrie (eds) *Critical Realism*. New York: Routledge.

BKK (2001) *Årsrapport 2000*. Bergen: *Bergenhalvøens kommunale kraftselskap* (BKK) AS.

BKK (2007) *Årsrapport 2006*. Bergen: BKK AS.

BNL (2004) *Næringpolitisk program for BNL—prioriterte saker 2004–2006*. Oslo: Byggenæringens landsforening (BNL).

BNL (2009) *Byggenæringen i tall*. Brochure. Oslo: Byggenæringens landsforening (BNL).

Boasson, E.L. (2005) *Klimaskapte beslutningsendringer? En analyse av klimahensyn i petroleumspolitiske beslutningsprosesser*. Masters thesis/FNI Report 13/2005, Lysaker, Norway: Fridtjof Nansen Institute.

Boasson. E.L (2013) *National Climate Policy Ambitiousness: A Comparative Study of Denmark, France, Germany, Norway, Sweden and the UK*. CICERO report 2013:02, Oslo.

Boasson, E.L. and Wettestad, J. (2013) *EU Climate Policy: Industry, Policy Interaction and External Environment*. Aldershot: Ashgate.

Boelhouwer, P. and van der Heijden, H. (1993) 'Housing policy in seven European countries', *Journal of Housing and the Built Environment*, 8(4):383–404.

Börzel, T. and Risse, T. (2003) 'Conceptualizing the domestic impact of Europe', in K. Featherstone and C.M. Radaelli (eds), *The Politics of Europeanisation*. Oxford: Oxford University Press.

Börzel, T. and Risse, T. (2006) 'Europeanization: the domestic impact of European Union politics', in K.E. Jørgensen, M. Pollack and B. Rosamond (eds), *Handbook of European Union Politics*. Thousand Oaks, CA: Sage.

Bourdieu, P. (1992) 'The practice of reflexive sociology', in P. Bourdieu and L.J.D. Wacquant, *An Invitation to Reflexive Sociology*. Cambridge: Polity Press.

Bourdieu, P. (2005) 'Principles of an economic sociology', in N.J. Smelser and R. Sweberg (eds), *Handbook of Economic Sociology*. Princeton, NJ: Princeton University Press.

Bourdieu, P. and Wacquant, L.J.D. (1992) *An Invitation to Reflexive Sociology*. Cambridge: Polity Press.

BP (2009) Capturing carbon dioxide. www.bp.com/sectiongenericarticle.do?categor yId=9023211&contentId=7043026 (Accessed 5 February 2010).

Budsjett-innst. S. nr. 9 (2000–2001) *Innstilling fra energi- og miljøkomiteen om bevilgninger på statsbudsjettet for 2001 vedkommende OED og MD*. Oslo: Stortinget, Energi- og miljøkomiteen.

Bulkley, H. and Moser, S.C (2007) 'Responding to climate change: governance and social action beyond Kyoto', introduction to special issue, *Global Environmental Politics* 7(2):1–10.

Bulmer, S. and Lequesne, C. (eds) (2005) *The Member States of the European Union*. Oxford: Oxford University Press.

CEN (European Committee for Standardization) (2009) Energy performance of buildings. Available at: www.cen.eu/cenorm/sectors/sectors/construction/sustainableconstruction/ epbd.asp (Accessed 21 January 2011).

Christensen, T. (2003) 'Narratives of Norwegian governance: elaborating the strong state tradition', *Public Administration* 81(1):163–90.

Christensen, T. and Lægreid, P. (2001) *New Public Management: The Transformation of Ideas and Practice*. Aldershot: Ashgate.

Christensen, T. and Lægreid, P. (2002) *Reformer og lederskap: omstilling i den utøvende makt*. Oslo: Universitetsforlaget.

Christensen, T., Lægreid, P., Roness, P. and Røvik, K.A. (eds) (2009). *Organisasjonsteori for offentlig sektor: instrument, kultur og myte*. Oslo: Universitetsforlaget

Christoff, P. and Eckersley, R. (2011) 'Comparing state responses', in J.S. Dryzek, R.B Norgaard and D. Schlosberg (eds), *Oxford Handbook of Climate Change and Society*. Oxford: Oxford University Press.

Clemens, E.S. and Cook, J.M. (1999) 'Politics and institutionalism: explaining durability and change', *Annual Review of Sociology*, 25:441–66.

Coen, D., Grant, W. and Wilson, G. 2010. 'Political science: perspectives on business and management', pp. 9–34 in D. Coen, W. Grant and G. Wilson (eds), *Oxford Handbook of Business and Government*. Oxford: Oxford University Press.

Cohen, M.D., March, J.G. and Olsen, J.P. (1972) 'A garbage can model of organizational choice', *Administrative Science Quarterly*, 17(1):1–25.

Cohen, M.D., March, J.G. and Olsen, J.P. (1979) 'People, problems, solutions and the ambiguity of relevance', in J.G. March and J.P. Olsen (eds), *Ambiguity and Choice in Organizations*. Oslo: Scandinavian University Press.

Collier, D. (1993) 'The comparative method', in A.W. Finifter (ed.) *Political Science: The State of the Discipline II*. Washington, DC: American Political Science Association.

Collier, D. and Gerring, J. (2009) *Concepts and Method in Social Science: The Tradition of Giovanni Sartori*. London: Routledge.

Collier, D.J., Laporte, J. and Seawright, J. (2008) 'Typologies: forming concepts and creating categorical variables', in J.M. Box-Steffensmeier, H.E. Brady and D. Collier (eds), *Oxford Handbook of Political Methodology*. Oxford: Oxford University Press.

Commission (2002) *ENERGY: Let Us Overcome Our Dependence*. Brussels: European Commission.

Commission (2004) *The Share of Renewable Energy in the EU*. Communication from the Commission COM 366. Brussels: European Commission.

Commission (2005a). *The Support of Electricity from Renewable Energy Sources*. Communication from the Commission COM 627. Brussels: European Commission.

Commission (2005b), *Communication from the Commission to the Council, the European Parliament, the European Economic and Social Committee and the Committee of the Regions—Winning the Battle Against Global Climate Change*. 9 February. COM(2005) 35 final. Brussels: European Commission.

Commission (2005c) *Green Paper: Energy Efficiency or Doing More With Less*. 22 June. COM (2005) 265 final. Brussels: European Commission.

Commission (2007a) *Limiting Global Climate Change to 2 degrees Celsius*. Communication. 10 January COM(2007) 2 final. Brussels: European Commission.

Commission (2007b) *Towards an Improved Policy on Industry Emissions*. Communication from the Commission to the Council, the European Parliament, the Economic and Social Committee and the Committee of the Regions. 21 December. Brussels: European Commission.

Commission (2008a) *The Support of Electricity from Renewable Energy Sources*. Commission staff working document, SEC 57. Brussels: European Commission.

Commission (2008b) *Proposal for a Directive of the European Parliament and of the Council on the Promotion of the Use of Energy from Renewable Sources*. Communication from the Commission. 23 January 19 final. 2008/0016 COD. Brussels: European Commission.

Commission (2009) *Synthesis of the Complete Assessment of all 27 National Energy Efficiency Action Plans as Required by Directive 2006/32/EC on Energy End-use Efficiency and Energy Services*. Commission Staff Working Document. 25 June. 11392/09. Brussels: European Commission.

Community Guidelines (2001) 'Community guidelines on state aid for environmental protection', *Official Journal of the European Communities* (2001/C 37/03).

Community Guidelines (2008) 'Community guidelines on state aid for environmental protection', *Official Journal of the European Union* (2008/C 82/01).

Compston, H. and I. Bailey (2012). *Climate Clever: How Governments Can Tackle Climate Change (and Still Win Elections)*. Abingdon: Routledge.

Council of the European Union (2007) Brussels European Council 8/9 March. Presidency Conclusions, 2 May. 7224/107.

Cowles, M.G., Caporaso, J. and Risse, T. (eds) (2001) *Transforming Europe: Europeanization and Domestic Change*. Ithaca, NY: Cornell University Press.

Cowles. M.G. and Risse, T. (2001) 'Transforming Europe', in M.G. Cowles, J. Caporaso and T. Risse (eds), *Transforming Europe: Europeanization and Domestic Change*. Ithaca, NY: Cornell University Press.

Culpepper, P. (2011). *Quiet Politics and Business Power: Corporate Control in Europe and Japan*. Cambridge: Cambridge University Press.

Czarniawska, B. and Sevón, G. (eds) (1996) *Translating Organizational Change*. Berlin: de Gruyter.

Dahl, R.A. (1961) *Who Governs? Democracy and Power in an American City*. New Haven, CT: Yale University Press.

Dahl, R.A. and Lindblom, C.E. (1953) *Politics, Economics and Welfare*. New York: Harper and Row.

Danermark, B., Ekström, M, Jakobsen, L. and Karlsson, J.C. (2002) *Explaining Society*. London: Routledge.

Davis, G.F., McAdam, D., Scott, W.R. and Zald, M.N. (eds) (2005) *Social Movements and Organization Theory*. Cambridge: Cambridge University Press.

Davis, J. (2006) '"And then there were Four . . ." A thumbnail history of oil industry restructuring, 1971–2005', in J.D. Davis (ed.), *The Changing World of Oil: An Analysis of Corporate Change and Adaptation*. Aldershot: Ashgate.

Delmas M. A. and Young, O. (2009) 'New perspectives on governance for sustainable development', in M.A. Delmas and O. Young (eds), *Governance for the Environment: New Perspectives*, pp. 3–11. New York: Cambridge University Press.

DiMaggio, P.J. (1988) 'Interests and agency in neo-institutional theory', in L.G. Zucker (ed.) *Institutional Patterns and Organizations: Culture and Environment*. Cambridge, MA: Ballinger.

DiMaggio, P.J. and Powell, W.W. (1991) *The Neo-institutionalism in Organizational Analysis*. Chicago, IL: University of Chicago Press.

DiMaggio, P.J. and Powell, W.W. ([1983] 1991) 'The iron cage revisited', in W.W. Powell and P.J. DiMaggio *The Neo-institutionalism in Organizational Analysis*. Chicago, IL: University of Chicago Press.

Directive (1996) 'Council Directive 96/61/EC of 24 September 1996 concerning integrated pollution prevention and control', *Official Journal* L 257:26–40.

Directive (2001) 'Directive 2001/77/EC of the European Parliament and of the Council on the promotion of electricity produced from renewable energy sources in the internal electricity market', *Official Journal* L 283:33–39.

Directive (2002) 'Directive 2002/91/EC of the European Parliament and of the Council of 16 December 2002 on the energy performance of buildings', *Official Journal* L 1:65–71.

Directive (2003) 'Directive 2003/87/EC of the European Parliament and of the Council of 13 October 2003 establishing a scheme for greenhouse gas emission allowance trading within the Community and amending Council Directive 96/61/EU', *Official Journal* L 275:32–46.

Directive (2009a) 'Directive 2009/31/EC of the European Parliament and of the Council of 23 April 2009 on the geological storage of carbon dioxide and amending Council Directive 85/337/EEC, European Parliament and Council Directives 2000/60/EC, 2001/80/EC, 2004/35/EC, 2006/12/EC, 2008/1/EC and Regulation (EC) No 1013/2006', *Official Journal* L 140:114–35.

Directive (2009b) 'Directive 2009/28/EC of the European Parliament and of the Council of 23 April 2009 on the promotion of the use of energy from renewable sources and amending and subsequently repealing Directives 2001/77/EC and 2003/30/EC', *Official Journal* L 140:16–62.

Directive (2010) 'Directive 2010/31/EC of the European Parliament and of the Council of 19 May 2010 on the energy performance of buildings (recast)', *Official Journal* L 153:13–35.

Dokka, T.H. and Hermstad, K. (2006) *Energieffektive boliger for fremtiden.* SNC 28/ECBCS Annex 38: Sustainable Solar Housing. Trondheim: SINTEFF Byggforsk.

Dorado, S. (2005) 'Institutional entrepreneurship, partaking, and convening', *Organization Studies*, 26(3):385–414.

Downs, A. (1972) 'Up and down with ecology—the "issue-attention cycle"'. *Public Interest*, 28 (Summer):38–50.

Dubash, N.K., M. Hagemann, N. Höne and P. Upadhyaya (2013) 'Developments in national climate change mitigation legislation and strategy', *Climate Policy*, 13(6):649–64.

Dubois, A. and Gadde, L-E. (2002) 'The construction industry as a loosely coupled system', *Construction Management and Economics*, 20(7):621–32.

Edelman, L. (2007) 'Overlapping fields and constructed legalities: the "endogeneity" of law', in J. O'Brien (ed.), *Private Equity, Corporate Governance, and the Dynamics of Capital Market Regulation*. London: Imperial College Press.

EFTA Court (2009) Judgment of the Court, 13 May 2009 (Failure by a contracting party to fulfil its obligations—Directive 2002/91/EC on the energy performance of buildings).

Egeberg, M. (2003) 'How bureaucratic structure matters: an organizational perspective', pp. 116–26 in B.G. Peters and J. Pierre (eds), *Handbook in Public Administration*. Thousand Oaks, CA: Sage.

Egeberg. M., Olsen, J.P. and Sæthren, H. (1978) 'Organisasjonssamfunnet og den segmenterte stat', in J.P. Olsen (ed.) *Politisk organisering*. Bergen: Universitetsforlaget.

Eidsiva (2007) *Årsrapport 2006*. Hamar, Norway: Eidsiva Energi AS.

Eising, R. (2004) 'Multilevel governance and business interests in the European Union', *Governance*, 17(2):211–45.

Energiloven (1990) [The Norwegian Energy Law] Lov om produksjon, omforming, overføring, omsetning, fordeling og bruk av energi m.m. (energiloven). Introduced 15 June 1990, last amended 1 January 2010.

Energimyndigheten (2005) *Konsekvenserna av en utvigad elcertifikatmarknad.* Eskilstuna, Sweden: Energimyndigheten.

ENI (2009) CCS: Carbon Dioxide (CO_2) Capture and Storage. Available at: www.eni.com/attachments/innovazione-tecnologia/technological-answers/scheda-cattura-sequestrazione-co$_2$-eng.pdf (Accessed 5 February 2010).

Enova (2006) *Resultat rapport 2005*. Trondheim: Enova.

Enova (2007) ESA krever tilbaketrekking av støtte. Press Release, 30 March.

Enova (2009) *Resultat og aktivitetsrapport for 2008*. Trondheim: Enova.

EnR (2008) *Implementation of the EU Energy Performance of Buildings Directive— A Snapshot Report*. European Energy Network.

EREC (2008) *European Renewable Energy Council*. Available at: www.erec.org (Accessed 24 January 2011).

ESA (2005a) Subject: The Energy Fund. Letter from the ESA to the Norwegian delegation to the European Union. 04.03.2005/1000–1. EFTA Surveillance Authority (ESA)

ESA (2005b) EFTA Surveillance Authority Decision of 18 May 2005.18.05.2005

ESA (2006). Subject: The Norwegian Energy Fund. Letter to the MPE. Case no. 57473. Even No. 365569. 03.05.2006.

ESA (2007a) Test Centre Mongstad. Letter from EFTA Surveillance Authority to the Norwegian Ministry of Government and Administrative Reform. Brussels, 21 August 2007.

ESA (2007b) Test Centre Mongstad. Brussels, Letter from EFTA Surveillance Authority to the Norwegian Ministry of Government and Administrative Reform. 31 October 2007.

ESA (2007c) Operation Aid for Electricity Production from Renewable Energy Sources. Letter to the MPE. 14 November 2007. Case no. 62684.

ESA (2008a) E-mail from Raphael Meyer, officer at the ESA, to Johan Vetlesen, Director, Ministry of Petroleum and Energy. 14 April 2008.

ESA (2008b) Norway Taken to the Court for Failing to Implement Directive on Energy Performance of Buildings. Press Release, 12 November.

ESA (2008c) EFTA Surveillance Authority Decision of 16 July 2008 on Test Centre Mongstad. Decision no. 503/08/COL.

ESA (2009) Notification of Carbon Capture Storage Project at Kårstø. Letter from EFTA Surveillance Authority to the Norwegian Ministry of Government and Administrative Reform. Brussels, 29 January 2009.

Farsund, A.A. (2000) 'Marked eller miljø?' *Tidsskrift for samfunnsforskning,* 41(3): 436–57.

Featherstone, K. (2003) 'Introduction', in K. Featherstone and C.M. Radaelli (eds), *The Politics of Europeanization.* Oxford: Oxford University Press.

Featherstone, K. and C.M. Radaelli (eds) (2003), *The Politics of Europeanization.* Oxford: Oxford University Press.

Finnemore, M. and Sikkink, K. (1998) 'International norm dynamics and political change', *International Organization,* 52:887–917.

Fligstein, N. (1997) 'Social skill and institutional theory', *American Behavioral Scientist,* 40(4):397–405.

Fligstein, N. (2001a) 'Social skill and the theory of fields', *Sociological Theory,* 19(2):105–25.

Fligstein, N. (2001b) *The Architecture of Markets.* Princeton, NJ: Princeton University Press.

Fligstein, N. (2008) *Euroclash.* Oxford: Oxford University Press.

Fligstein, N. and Mara-Drita, I. (1996) 'How to make a market: reflections on the attempt to create a Single Market in the European Union', *American Journal of Sociology,* 102(1):1–33.

Fligstein, N. and McAdam, D. (2012) *A Theory of Fields.* Oxford: Oxford University Press.

Foreningen Næringseiendom (2009) Medlemsbedriftene [list of members]. Available at: www.foreningen-nar ingseiendom.no/medlemsbedriftene/ (Accessed 30 September 2009).

Fouquet, D. and Johansson, T.B. (2008) 'European renewable energy policy at crossroads', *Energy Policy,* 36:4079–92.

Friedland, R. and Alford, R.R. (1991) 'Bringing society back in: symbols, practices and institutional contradictions', in W.W. Powell and P. J. DiMaggio (eds) *The Neo-institutionalism in Organizational Analysis.* Chicago, IL: University of Chicago Press.

Fuglseth, B.B. (2009) *Regulative Change Targeting Energy Performance of Buildings in Sweden: Key Drivers and Main Implications.* FNI Report 2/2009. Lysaker, Norway: Fridtjof Nansen Institute.

Gais, T.L, Peterson, M.A. and Walker, J.L. (1984) 'Interest groups, iron triangles and representative institutions in American national government', *British Journal of Political Science,* 14(2):161–85.

Gallagher, M., Laver, M. and Mair, P. (2011) *Representative Government in Modern Europe,* 5th edn. Maidenhead: McGraw-Hill Higher Education.

Galtung, J. (1971) 'A structural theory of imperialism', *Journal of Peace Research,* 8:81–117.

Garud, R., Hardy, C. and Maguire, S. (2007) 'Institutional entrepreneurship as embedded agency: an introduction to the special issue', *Organization Studies*, 28(7):957–69.

Gassco (2009) *Naturkraft Integration at Kårstø. Mapping Study Report*. Kopervik, Norway: Gassco.

Gassco and Gassnova (2010) *Kårstø Integration Pre-Feasibility Study*. Porsgrunn, Norway: Gassnova.

George, A.L. and Bennett, A. (2005) *Case Studies and Theory Development*. Cambridge, MA: MIT Press.

Gerring, J. (2008) 'Case selection for case-study analysis: qualitative and quantitative techniques', in J.M. Box-Steffensmeier, H.E. Brady and D. Collier (eds), *Oxford Handbook of Political Methodology*. Oxford: Oxford University Press.

Giddens, A. (2009) *Politics of Climate Change*. Cambridge: Polity Press.

Gjerløw, T. (1996) *17 år med ENØK og alternative energikilder*. Oslo: Olje- og energidepartementet [Ministry of Petroleum and Energy].

Gjerset, M. (2007) *Innspill i lys av NVE-rapporten og veien videre for regjeringens arbeid med fullskala rensing på Kårstø*. Oslo: ZERO.

Goffmann, E (1974) *Frame Analysis: An Essay in the Organization of Experience*. Cambridge, MA: Harvard University Press.

Goodin, R.E., Rein, M. and Moran, M. (2006) 'The public and its policies', in R.E. Goodin, M. Rein and M. Moran (eds), *Oxford Handbook of Public Policy*. Oxford: Oxford University Press.

Greenwood, R. and Suddaby, R. (2006) 'Institutional entrepreneurship in mature fields: the Big Five accounting firms', *Academy of Management Journal*, 49(1): 27–48.

Greenwood, R., Suddaby, R. and Hinings, C. R. (2002) 'Theorizing change: the role of professional associations in the transformation of institutional fields', *Academy of Management Journal*, 45(1):58–80.

Gross, N. (2009) 'A Pragmatist Theory of Social Mechanisms', *American Sociological Review* 74:358–79.

Gulick, L. (1937) 'Notes on the theory of organization', in L. Gulick and L. Urwick, *Papers on the Science of Administration*. Concord, NH: Rumford Press.

Gupta, S., Tirpak, D.A., Burger, N., Gupta, J., Höhne, A. et al. (2007) 'Policies, instruments and co-operative arrangements', in B. Metz, O.R. Davidson, P.R. Bosch, R. Dave and L.A. Meyer (eds), *Climate Change 2007: Mitigation*. Cambridge: Cambridge University Press.

Guttu, J. (2003) *Den gode boligen*. PhD thesis, Oslo School of Architecture and Design. Oslo: Unipub.

Hafslund (2007) *Årsrapport 2006*. Oslo.

Hall, P.A. (2003) 'Aligning ontology and methodology in comparative research', in J. Mahoney and D. Rueschemeyer (eds), *Comparative Historical Analysis in the Social Sciences*. Cambridge: Cambridge University Press.

Hall, P.A. and Soskice, D. (2001) 'Introduction', in P.A. Hall and D. Soskice (eds), *Varieties of Capitalism: The Institutional Foundation of Comparative Advantage*. Oxford: Oxford University Press.

Halse, A. (2005) *Passive Houses in Norway*, report, Centre for Technology, Innovation and Culture. Oslo: University of Oslo.

Hancké, B. (ed.) (2009) *Debating Varieties of Capitalism*. Oxford: Oxford University Press.

Hanisch, T.J. and Nerheim, G. (1992) *Fra vantro til overmot?* Oslo: Leseselskapet.

Hardy, C. and Maguire, S. (2008) 'Institutional entrepreneurship', in R. Greenwood, C. Oliver, K. Sahlin and R. Suddaby (eds), *SAGE Handbook of Organizational Institutionalism*. Thousand Oaks, CA: Sage.

Harris, P.G. (ed.) (2007a) *Europe and Global Climate Change: Politics, Foreign Policy and Regional Development*. Cheltenham: Edward Elgar.

Harris, P.G. (2007b) 'Europe and the politics and foreign policy of global climate change', in P.G. Harris (ed.) *Europe and Global Climate Change: Politics, Foreign Policy and Regional Development*. Cheltenham: Edward Elgar.

Harris, P.G. (2007c) 'Explaining European responses to global climate change: power, interests and ideas on domestic and international politics', in P.G. Harris (ed.), *Europe and Global Climate Change: Politics, Foreign Policy and Regional Development*. Cheltenham: Edward Elgar.

Harrison, K. and L. M. Sundstrom (ed.) (2010). *Global Commons, Domestic Decisions*. Cambridge, MA: MIT Press.

Haverland, M. (2000) 'National adaptation to European integration: the importance of institutional veto points', *Journal of Public Policy*, 20(1):83–103.

Havskjold, M. and Halseth, A. (2007) *Fornybar varme 2020*. Oslo: Xriga.

Heclo, H. (1978) 'Issue networks and executive establishment', in A. King and S.H. Beer (eds), *The New American Political System*. Washington, DC: American Enterprise Institute.

Hedström, P. (2008) 'Studying mechanisms to strengthen causal inferences in quantitative research', in J.M. Box-Steffensmeier, H.E. Brady and D. Collier (eds), *Oxford Handbook of Political Methodology*. Oxford: Oxford University Press.

Hernes, G. (1983) *Det moderne Norge: makt og styring*. Oslo: Gyldendal Norsk Forlag.

Hoffmann, A. ([1997] 2001) *From Heresy to Dogma*, 2nd edn. Stanford, CA: Stanford University Press.

Hoffmann, M. (2011) *Climate Governance at the Crossroads*. Oxford: Oxford University Press.

Holm, P. (1995) 'The dynamics of institutionalization: transformation processes in Norwegian fisheries', *Administrative Science Quarterly*, 40:398–422.

Hood, C. (2007) 'Intellectual obsolesence and intellectual makeovers: reflections on the tools of government after two decades', *Governance: An International Journal of Policy, Administration, and Institution*, 20(1):127–44.

Hooghe, L. (2001) *The European Commission and the Integration of Europe: Images of Europe*. Cambridge: Cambridge University Press.

Hooghe, L. and Marks, G. (2001) *Multilevel Governance and European Integration*. Oxford: Rowman and Littlefield.

Hubak, M. (1998) *Synlig kostnad—skjult gevinst*. Dr.Polit dissertation. Trondheim: Senter for teknologi og samfunn, report No. 39.

Husbanken (1995) *The Growth of Good Housing*. Brochure. Oslo: Husbanken.

Husbanken (2003) *Husbankhus halverer energibehovet*. Booklet. Oslo: Husbanken.

Husbanken (2006) Husbanken finansierer flere lavenergiboliger. Press Release, 10 September.

Innst. S. nr. 122 (1999–2000) *Innstilling fra energi—og miljøkomiteen om energipolitikken, St.meld. nr. 29 (1998–1999), unntatt kap. 8 om Kraftkontrakter med industrien; og forslag oversendt fra Stortinget 13. oktober 1999*. Oslo: Stortinget.

Innst. S. nr. 167 (2002–2003) *Innstilling fra energi—og miljøkomiteen om innenlands bruk av naturgass mv.* Oslo: Stortinget.

Innst. S. nr. 205 (2006–2007) *Innstilling fra energi—og miljøkomiteen om samarbeid om håndtering av CO$_2$ på Mongstad*. Oslo: Stortinget.

Innst. S. nr. 145 (2007–2008) *Innstilling fra energi—og miljøkomiteen om norsk klimapolitikk*. Oslo: Stortinget.

Innst. S. nr. 206 (2008–2009) *Innstilling fra energi—og miljøkomiteen om investering i teknologisenter for CO$_2$-håndtering på Mongstad*. Oslo: Stortinget.

Jakobsen, V.E., Hauge, F., Holm, M. and Kristiansen, B. (2005) *CO$_2$ til EOR på norsk sokkel—en mulighetsstudie*. Oslo: Bellona.

Jensen, O.M. (2009) *Virkemidler til fremme af energisparelser i bygninger.* SBi 2009:06. Ålborg, Denmark: Statens Byggforskningsinstitutt [Danish Building Research Institute].

Jensen, O.M., Wittchen, K.B. and Thomsen, K.E. (2009) *Towards Very Low Energy Buildings.* Ålborg, Denmark: Danish Building Research Institute.

Johnstad, T. (1993) *En konkurransedyktig byggenæring.* SNF-rapport 27. Bergen: Stiftelsen for samfunns- og næringslivsforskning.

Jordan, A. and Schout, A. (2006) *The Coordination of the European Union: Exploring the Capacities of Networked Governance.* Oxford: Oxford University Press.

Jordan, A. and Huitema, D. (forthcoming 2014) 'Policy Innovation in a Changing Climate', introduction submitted for a special issue of *Global Environmental Change.*

Jordan, A., Huitema, D., van Asselt, H., Rayner T. and Berkhout, F. (eds) (2010), *Climate Change Policy in the European Union: Confronting the Dilemmas of Mitigation and Adaptation.* Cambridge: Cambridge University Press.

Jordan, A. and Liefferink, D. (2005) *Environmental Policy In Europe. The Europeanisation of National Environmental Policy.* Abingdon: Routledge.

Kallestrup, M. (2005) *Europæisering af nationalstaten.* Copenhagen: Jurist- og økonomforbundets forlag.

Kemeny, J. (2001) 'Comparative housing and welfare', *Journal of Housing and the Built Environment,* 16:53–70.

Keulenaer, H. and Gerwen, R. (2008) *The Passive House in the Electricity System of the Future.* Brussels: European Copper Institute.

King, G., Keohane, R.O. and Verba, S. (1994) *Designing Social Inquiry: Scientific Inference and Qualitative Research.* Princeton, NJ: Princeton University Press.

Kingdon, J.W. ([1984] 2011) *Agendas, Alternatives, and Public Policies.* Boston, MA: Little, Brown.

Klein, R. and Marmor, T. R. (2006) 'Reflections on policy analysis: putting it together again', pp. 892–912 in R.E. Goodin, M. Rein and M. Moran (eds), *Oxford Handbook of Public Policy.* Oxford: Oxford University Press.

Kohler-Koch, B. (1999) 'The evolution and transformation of European governance', pp. 14–35 in B. Kohler-Koch and R. Eising (eds), *The Transformation of Governance in the European Union.* London: Routledge.

KRD [Ministry of Local Government and Regional Development] (2005) *Miljøhandlingsplan 2005–2008.* Oslo: Ministry of Local Government and Regional Development.

KRD [Ministry of Local Government and Regional Development] (2006a) Høringsforslag juni 2006. Available at: www.regjeringen.no/upload/kilde/krd/hdk/2006/0004/ddd/pd fv/283635-horingsnotat_tek.pdf (Accessed 24 February 2007).

KRD [Ministry of Local Government and Regional Development] (2006b) Høring— endringer i tekniske forskrifter til plan- og bygningsloven (TEK) og forskrift om saksbehandling og kontroll (SAK). Høringsinnspill. Available at: www.regjeringen. no/nb/dep/ krd/dok/hoeringer/hoeringsdok/2006/horing-endringer-i-tekniske-forskrifter-.html?id=98401 (Accessed 24 February 2007).

KRD [Ministry of Local Government and Regional Development] (2007) *Forskrift om endringer i forskrift 22 January 1997. Nr. 33 til Plan og Bygningsloven om krav til byggverk og produkter til byggverk (TEK). Fastsatt av Kommunal- og regionaldepartementet den 26. januar 2007, med hjemmel i plan- og bygningsloven av 14. juni 1985 nr. 77 §§ 6, 77, 81, 82, 84.* Oslo: Ministry of Local Government and Regional Development.

KRD [Ministry of Local Government and Regional Development] (2009) *Bygg for framtida. Miljøhandlingsplan for bolig- og byggsektoren 2009–2012.* Oslo: Ministry of Local Government and Regional Development.

KRD [Ministry of Local Government and Regional Development] (2010) *Forskrift om endring i forskrift 26. mars 2010 nr. 489 om tekniske krav til byggverk (byggteknisk forskrift)*. Oslo: Ministry of Local Government and Regional Development.

Kuhn, T. (2001) 'Implications of the "Preussen Elektra"', *Legal Issues of Economic Integration*, 28(3):361–76.

Lachapelle, E. and Paterson, M. (2013) 'Drivers of national climate policy', *Climate Policy*, 13(5):547–71.

Lascoumes, P. and Le Gales, P. (2007) 'Introduction : Understanding public policy through Its instruments—from the nature of instruments to the sociology of public policy instrumentation', *Governance: An International Journal of Policy, Administration, and Institution*, 20(1):1–21

Lavenergiboliger (2009) 'Den mest miljøvennlige energien er den som ikke blir brukt'. Available at: www.lavenergiboliger.no (Accessed 4 April 2009).

Lawrence, T.B. and Suddaby, R. (2006) 'Institutions and institutional work', pp. 215–54 in S.R. Clegg, C. Hardy, T.B. Lawrence and W.R. Nord (eds), *The SAGE Handbook of Organization Studies*, 2nd edn. London: Sage.

Lawrence, T.B., Suddaby, R. and Leca, B. (2009) 'Introduction: theorizing and studying institutional work', pp. 1–27 in Lawrence, T.B., Suddaby, R. and Leca, B. (eds) *Institutional Work: Actors and Agency in Institutional Studies of Organizations*. Cambridge: Cambridge University Press.

Leblebici H., Salancik, G.R., Copay, A. and King, T. (1991) 'Institutional change and the transformation of interorganizational fields: an organizational history of the U.S. radio broadcasting industry', *Administrative Science Quarterly*, 36:333–63.

Leca, B., Battilana, J. and Boxenbaum, E. (2006) 'Taking stock of institutional entrepreneurship: What do we know? Where do we go?' Paper presented at the Academy of Management Meetings, 11–16 August 2006, Atlanta, GA.

Lie, E. (2005) *Oljerikdommer og internasjonal ekspansjon*. Oslo: Pax.

Lieberson, S. (1971) 'An empirical study of military–industrial linkages', *American Journal of Sociology*, 76(4):562–84.

Lijphart, A. (1971) 'Comparative politics and comparative method', *American Political Science Review*, 65(3):682–93.

Lindblom, C. (1977) *Politics and Markets: The World's Political-Economic Systems*. New York: Basic Books.

LO [The Norwegian Confederation of Trade Unions] (2001) *Ta naturgassen i bruk!* Oslo: LO.

Lounsbury, M. (2001) 'Institutional sources of practice variation: Staffing college and university recycling programs', *Administrative Science Quarterly*, 46(1):29–56.

Lounsbury, M. (2007) 'A tale of two cities', *Academy of Management Journal*, 50(2):289–307.

Lounsbury, M., Ventresca, M. and Hirsch, P. M. (2003) 'Social movements, field frames and industry emergence: a cultural–political perspective on US recycling', *Socio-Economic Review*, 1:71–104.

Lounsbury, M. and Crumley, E.T. (2007) 'New practice creation: an institutional perspective on innovation', *Organization Studies*, 28(7):993–1012.

Lowi, T.J. (1964) 'Review: American business, public policy, case studies and political theory', *World Politics*, 16(4):677–715.

Lowi, T.J. (1969) *The End of Liberalism: Ideology, Policy and the Crisis of Public Authority*. New York: W.W. Norton.

Lukes, S. (1974) *Power: A Radical View*. London: Macmillan.

Maguire, S., Hardy, C. and Lawrence, T.B. (2004) 'Institutional entrepreneurship in emerging fields: HIV/AIDS treatment advocacy in Canada', *Academy of Management Journal*, 47(5):657–79.

Mahoney, J. (2003) 'Strategies of causal assessment in comparative historical analysis', pp. 337–72 in J. Mahoney and D. Rueschemeyer (eds), *Comparative Historical Analysis in the Social Sciences*. Cambridge: Cambridge University Press.

Mahoney, J. and Rueschemeyer, D. (2003) 'Comparative historical analysis: achievements and agendas', pp. 3–38 in J. Mahoney and D. Rueschemeyer (eds), *Comparative Historical Analysis in the Social Sciences*. Cambridge: Cambridge University Press.

Mahoney, J. and Goertz, G. (2006) 'A tale of cultures: contrasting quantitative and qualitative research', *Political Analysis*, 14:227–49.

Mahoney, J. and Thelen, K. (2010) 'A theory of gradual institutional change', in J. Mahoney and K. Thelen (eds), *Explaining Institutional Change*. Cambridge: Cambridge University Press.

Majone, G. (2006) 'Agenda setting', in R.E. Goodin, M. Rein and M. Moran (eds), *Oxford Handbook of Public Policy*. Oxford: Oxford University Press.

March, J.G. and Olsen, J.P. (1983) 'Organizing political life: what administrative reorganization tells us about government', *American Political Science Review*, 77(2):281–96.

March, J.G. and Olsen, J.P. (1989) *Rediscovering Institutions*. New York: Free Press.

March, J.G. and Olsen, J.P. (1998) 'The institutional dynamics of international political orders', *International Organization*, 52(4):943–69.

Marks, G., Hooghe, L. and Blank, K. (1996) 'European integration from the 1980s: state-centric v. multi-level governance', *Journal of Common Market Studies*, 34(3):341–78.

Mastenbroek, E. (2005) 'EU compliance: still a black hole?' *Journal of European Public Policy*, 12(6):1103–20.

Mazey, S. and Richardson, J. (2006) 'Interest groups and EU policy-making: organisational logic and venue shopping', pp. 247–68 in J. Richardson (ed.) *European Union: Power and Policy-making*. London: Routledge.

Metz, B., Davidson, O.R., Bosch, P.R., Dave, R. and Meyer, L.A. (eds) (2007) *Climate Change 2007: Mitigation*. Working Group III contribution to the Fourth Assessment Report of the Intergovernmental Panel on Climate Change. Cambridge: Cambridge University Press.

Meyer, J.W. (2000) 'Globalization: sources and effects on national states and societies', *International Sociology*, 15(2):233–48.

Meyer, J.W., Boli, J., Thomas, G.M. and Ramirez, F.O. (1997) 'World society and the nation-state', *American Journal of Sociology*, 103(1):144–81.

Meyer, J.W. and Rowan, B. (1991) 'Institutionalized organizations: formal structure as myth and ceremony', pp. 41–62 in P.J. DiMaggio and W.W. Powell (eds), (1991) *The Neo-institutionalism in Organizational Analysis*. Chicago, IL: University of Chicago Press [first published in *American Journal of Sociology* 1983].

Midttun, A. (1987) *Segmentering, institusjonelt etterslep og industriell omstilling*. Doctoral dissertation. Uppsala, Sweden: Uppsala University.

Miles, E.L., Underdal, A., Andresen, S., Wettestad, J., Skjærseth, J.B. and Carlin, E.M. (2002) *Environmental Regime Effectiveness: Confronting Theory with Evidence*. Cambridge, MA: MIT Press.

Ministry of Finance (2009) 'Statens inntekter og utgifter—en oversiktstabell'. Press release 13 October, Oslo.

Minogue, M. ([1983] 1993) 'Theory and practice in public policy and administration' pp. 10–29 in Hill, M. (ed.) *The Policy Process: A Reader*. Hemel Hempstead: Prentice Hall/Harverster Wheatsheaf. [First published in *Policy and Politics* 1983]

Mintrom, M. (1997) 'Policy entrepreneurs and the diffusion of innovation', *American Journal of Political Science*, 41(3):738–70.

Moravcsik, A. (1998) *The Choice for Europe: Social Purpose & State Power from Messina to Maastricht*. Ithaca, NY: Cornell University Press.

Moravcsik. A. (1999) 'A new statecraft? Supranational entrepreneurs and international cooperation', *International Organization*, 53(2):267–306.

Moravcsik, A. and Schimmelfennig, F. (2009) 'Liberal intergovernmentalism', pp. 67–87 in T. Diez and A. Wiener (eds), *European Integration Theory*. Oxford: Oxford University Press.

Mörth, U. (2003) 'Europeanization as interpretation, translation and editing of public policies', pp. 159–75 in K. Featherstone and C.M. Radaelli (eds), *The Politics of Europeanization*. Oxford: Oxford University Press.

MPE [Ministry of Petroleum and Energy] (2002) *Avtale mellom Den norske stat v/ Olje-og energidepartementet og Enova SF*. Oslo: Ministry of Petroleum and Energy.

MPE [Ministry of Petroleum and Energy] (2003) Regard notification of the Energy Fund—request for additional information. Letter from the MPE to ESA. MPE 2003/00000. 11.09.2003.

MPE [Ministry of Petroleum and Energy] (2004) Regard notification of the Energy Fund—request for additional information. Letter from the MPE to ESA MPE 2002/1281. 09.06.2004.

MPE [Ministry of Petroleum and Energy] (2005a) Høringsuttalelser [Responses to public consultations]. Available at: www.re gjeringen.no/nb/dep/oed/dok/hoeringer/hoeringsdok/2005 (Accessed 27 January 2011).

MPE [Ministry of Petroleum and Energy] (2005b) *Tildelingsbrev til Gassnova*. Statsbudsjettet 2005.

MPE [Ministry of Petroleum and Energy] (2006a) Støtteordning for fornybarelektrisitet. Press release, 5 October.

MPE [Ministry of Petroleum and Energy] (2006b) *Samarbeid om håndtering av CO_2 på Mongstad ('Gjennomføringsavtalen')*. *Avtale mellom Staten v/Olje- og energidepartementet og Statoil ASA*. Oslo: Ministry of Petroleum and Energy.

MPE [Ministry of Petroleum and Energy] (2007a) *Avtale mellom Den norske stat v/Olje- og energidepartementet og Enova SF*. Oslo: Ministry of Petroleum and Energy.

MPE [Ministry of Petroleum and Energy] (2007b) Hørings-utkast til forskrift om støtteordning for produksjon av elektrisk energi fra fornybare energikilder. Available at: www. regjeringen.no/nb/dep/oed/dok/hoeringer/hoeringsdok/2007 (Accessed 2 October 2007).

MPE [Ministry of Petroleum and Energy] (2007c) Notification—Test Centre Mongstad. Letter to EFTA Surveillance Authority. 06/01763–18. 4 July 2007.

MPE [Ministry of Petroleum and Energy] (2007d) Notification of Test Centre Mongstad—reply to the first information request. Letter to EFTA Surveillance Authority. 18 September 2007.

MPE [Ministry of Petroleum and Energy] (2007e) State aid—Test Centre Mongstad—reply to the third information request. Letter til EFTA Surveillance Authority. 29 November 2007.

MPE [Ministry of Petroleum and Energy] (2007f) Høring av endringer i energiloven (Energitilstand i bygninger) Available at: www.regjeringen.no/nb/dep/oed/dok/hoeringer/ hoeringsdok/2007/horing-av-endringer-i-energiloven—-ener. html?id=473162 (Accessed 12 May 2007).

MPE [Ministry of Petroleum and Energy] (2008a) *Fakta 2008 om energi og vannressurser i Norge*. Oslo: Ministry of Petroleum and Energy.

MPE [Ministry of Petroleum and Energy] (2008b) Notification of Test Centre Mongstad—reply to the second information request. Letter til EFTA Surveillance Authority. 18 January 2008.

MPE [Ministry of Petroleum and Energy] (2008c) Tildelingsbrev til Gassnova. Statsbudsjettet 2008. Oslo: Ministry of Petroleum and Energy.

MPE [Ministry of Petroleum and Energy] (2009a) Tildelingsbrev til Gassnova. Stats-budsjettet 2009. Oslo: Ministry of Petroleum and Energy.

MPE [Ministry of Petroleum and Energy] (2009b) *Fakta. Norsk Petroleumsverk-semd. 2009.* Oslo: Ministry of Petroleum and Energy.

MPE [Ministry of Petroleum and Energy] (2009c) *Fakta. Energi og vannressurser i Norge.* Oslo: Ministry of Petroleum and Energy.

MPE [Ministry of Petroleum and Energy] (2010a) *Avtale mellom Den norske stat ved Olje- og energidepartementet og Enova SF om forvaltning av midlene fra Energifondet i perioden 1. juni 2008 til 31. desember 2011.* Oslo: Ministry of Petroleum and Energy.

MPE [Ministry of Petroleum and Energy] (2010b) Møte om etableringen av et felles sertifikatmarked. Press release, 29 April.

MPE [Ministry of Petroleum and Energy] (2010c) Høringsnotat (forslag til lovvedtak) om lov om elsertifikater. 8 December 2010. Oslo: Ministry of Petroleum and Energy.

MPE [Ministry of Petroleum and Energy] (2010d) Full scale CCS at Mongstad. Press Release. 2 May 2010.

MPE [Ministry of Petroleum and Energy] (2010e) StatoilHydro ASA. Available at: www.regjeringen.no/nb/dep/oed/tema/statlig_engasjement_i_petroleumsvirk-somh/statoil-asa.html?id=444383 (Accessed 5 May 2010).

NCC (2008) *Historie.* Available at: www.ncc.no/no/OM-NCC/NCC-i-Norge/Historie/. (Accessed 20 August 2008).

Nerheim, G. (1996) *En gassnasjon blir til. Norsk Oljehistorie, bind 2.* Oslo: Lese-selskapet.

Newman, A.L. (2008) 'Building transnational civil liberties: transgovernmental entrepreneurs', *International Organization,* 62 (Winter):103–30.

Nilsen, Y. (2001) *En felles plattform? Norsk oljeindustri og klimadebatten i Norge frem til 1998.* PhD thesis. Oslo: University of Oslo.

Nilsen,Y. and Thue, L. (2006) *Statens kraft 1965–2006.* Oslo: Universitetsforlaget.

Nilsson, M., Nilsson, L. and Ericsson, K. (2008). *Rapid Turns in European Renewable Energy Policy.* FNI Report 9/2008. Lysaker, Norway: Fridtjof Nansen Institute.

Norgaard, K. M. (2011) 'Climate denial: emotion, psychology, culture, and political economy', pp. 399–423 in J.S. Dryzek, R.B Norgaard and D. Schlosberg (eds), *Oxford Handbook of Climate Change and Society.* Oxford: Oxford University Press.

NOU (2002) *Gassteknologi, miljø og verdiskaping. Utredning fra en ekspertgruppe oppnevnt av Olje- og energidepartementet 5. oktober 2001. Avgitt 1. mars 2002.* Utredning nr. 7. Oslo: Norsk Offentlig Utredning [Official Norwegian Report].

NTE (2004) *Årsrapport 2003.* Steinkjer, Norway: Nord Trøndelag Elektrisitetsverk.

NVE (2004) *Grønne sertifikater: Utredning om innføring av et pliktig sertifikat-marked for kraft fra fornybare energikilder.* Report no. 11/2004. Oslo: Norwegian Water Resources and Energy Directorate (NVE).

NVE (2006) CO_2-håndtering på Kårstø—Fangst, transport og lagring. Report no. 13/2006. Oslo: Norwegian Water Resources and Energy Directorate (NVE).

NVE (2010a) Anleggskonsesjon. Industrikraft Møre AS. Brev, 19 February 2010.

NVE (2010b) *Forskrift om energimerking av bygninger og energivurdering av tekniske anlegg. Forslag til endringer i forskrift av 18.12.2009 nr. 1665.* Oslo: Norwegian Water Resources and Energy Directorate NVE).

OECD (2007) *Glossary of Statistical Terms.* http://stats.oecd.org/glossary/detail. asp?ID=7214 (Accessed 10 April 2014).

Olsen, J.P. (1983) *Organized Democracy: Political Institutions in a Welfare State: the Case of Norway.* Oslo: Universitetsforlaget.

Olsen, J.P. (1992) 'Analyzing institutional dynamics', *Staatswissenschaften und Sta-atspraxis.* 2:247–71.

Olsen, J.P. (2006) *Europe in Search of Political Order: An Institutional Perspective on Unity/Diversity, Citizens/Their Helpers, Democratic Design/Historical Drift and Co-existence of Orders*. Oxford: Oxford University Press.

Olsen, J. P. (2007) Johan P. Olsen's response [to being appointed EGOS honorary member 2007] http://www.egosnet.org/egos/about_egos/honorary_members/2007_johan_p_olsen/johan_p_olsens_response (Accessed 15 July 2013).

Olsen, J.P. (2010) *Governing Through Institution Building. Institutional Theory and Recent European Experiments in Democratic Organization*. Oxford: Oxford University Press.

Osmundsen, P., Mohn, K., Misund, B. and Asche, F. (2006) 'Is oil supply choked by financial market pressures?' *Energy Policy*, 35:467–74.

Ot. prp. nr. 35 (2000–2001) *Om lov om endringar i lov 29. juni 1990 nr. 50 om produksjon, omforming, overføring, omsetning og fordeling av energi m.m. (energilova)*. Oslo: Ministry of Petroleum and Energy.

Ot. prp. nr. 13 (2004–2005) *Om lov om kvoteplikt og handle med kvoter for utslipp av klimagasser (klimakvoteloven)*. Oslo: Ministry of the Environment.

Ot. prp. nr. 24 (2008–2009) *Om lov om endringar i lov 29. Juni 1990 nr. 50 om produksjon, omforming, overføring, omsetning, fordeling og bruk av energi m.m. (energiloven)*. Oslo: Ministry of Petroleum and Energy.

Panitch, L. (1980) 'Recent theorization of corporatism: reflections on growth industry', *British Journal of Sociology*, 31(2):159–87.

Passivhuscentrum (2009) Marknaden for passivhus. Available at: www.passivhuscentrum.se/ marknaden.html (Accessed 4 April 2009).

Patomäki, H. and Wight, C. (2000) 'After postpositivism? The promises of critical realism', *International Studies Quarterly*, 44:213–37.

Peab (2008). *Årsredovisning 2007*. Förslöv, Sweden: Peab.

Pedersen, P.H. and Steen, Ø. (2005) *Skaperevne og initiativ*. Oslo: Anlegg Media.

Petroleumsloven (1996) [The Norwegian Petroleum Law] *Lov om petroleumsvirksomhet [petroleumsloven]*. LOV 1996–11–29 No. 72: Introduced 1 July 1996. Last amended 1 July 2010. Oslo: Ministry of Petroleum and Energy.

Pfeffer, J. (1981) *Power in Organizations*. Marshfield, MA: Pitman.

Pfeffer, J. and Salanick, G.R. (1978) *The External Control of Organizations: A Resource Dependence Perspective*. New York: Harper and Row.

Pierson, P. (2004) *Politics in Time: History, Institutions and Social Analysis*. Princeton, NJ: Princeton University Press.

Polsby, N.W. (1984) *Political Innovation in America: The Politics of Policy Initiation*. New Haven, CT: Yale University Press.

Rabe, B. (2008) 'States on steroids: the intergovernmental odyssey of American climate policy', *Review of Policy Research*, 25(2):105–28.

Radaelli, C. (2003) 'The Europeanization of public policy', pp. 27–56 in K. Featherstone and C. Radaelli (eds), *The Politics of Europeanization*. Oxford: Oxford University Press.

Ragin, C. (1987) *The Comparative Method: Moving Beyond Qualitative and Quantitative Strategies*. Berkeley, CA: University of California Press.

Rao, H., Monin, P. and Durand, R. (2003) 'Institutional change in Toque Ville: Nouvelle cuisine as an identity movement in French gastronomy', *American Journal of Sociology*, 108(4):795–843.

Reay, R. and Hinings, C.R. (2009) 'Managing the rivalry of competing institutional logics', *Organization Studies*, 30(6):629–52.

Reiersen, E. and Thue, E. (1996) *De tusen hjem. Den Norske Stats Husbank*. Oslo: AdNotam Gyldendal.

Reitan, M. (1998) *Interesser og institusjoner i miljøpolitikken*. PhD thesis. Oslo: University of Oslo, Department of Political Science.

Rhodes, R.A.W. (1997) *Understanding Governance: Policy Networks, Governance, Reflexivity and Accountability.* Buckingham: Open University Press.

Riksrevisjonen (2010a) *Riksrevisjonens undersøkelse av Enova SFs drift og forvaltning.* Dokument 3:6. Oslo: Office of the Auditor General of Norway (Riksrevisjonen).

Riksrevisjonen (2010b) *Riksrevisjonens undersøkelse av måloppnåelse i klimapolitikken.* Dokument 3:5. Oslo: Office of the Auditor General of Norway (Riksrevisjonen).

Risse, T., Cowles, M.G. and Caparaso, J. (2001) 'Europeanization and Domestic Change', pp. 1–20 in M.G. Cowles, J. Caporaso and T. Risse (eds), *Transforming Europe: Europeanization and Domestic Change.* Ithaca, NY: Cornell University Press.

Roberts, N. and King, P.J. (1991) 'Policy entrepreneurs: their activity structure and function in the policy process', *Journal of Public Administration Research and Theory,* 1(2):147–75.

Røvik, K.A. (1998) *Moderne organisasjoner: trender i organisasjonstenkningen ved tusenårskiftet.* Bergen: Fagbokforlaget.

Røvik, K.A. (2007) *Trender og translasjoner: ideer som former det 21. århundrets arbeidsliv.* Oslo: Universitetsforlaget.

Røvik, K.A. (2011) 'From fashion to virus: an alternative theory of organizations' handling of management ideas', *Organization Studies,* 32(5):631–53.

Rowlands, I.H. (2005)'The European Directive on Renewable Electricity: conflicts and compromises'. *Energy Policy,* 33(8):965–74.

Rueschemeyer, D. (2003) 'Can one or a few cases yield theoretical gains?', pp. 305–36 in J. Mahoney and D. Rueschemeyer (eds), *Comparative Historical Analysis in the Social Sciences.* Cambridge: Cambridge University Press.

Ryghaug, M. (2003) *Towards a Sustainable Aesthetics: Architects Constructing Energy Efficient Buildings.* Doctoral dissertation in political science, Report 62/03, Trondheim: Norwegian University of Science and Technology (NTNU), Department of Interdisciplinary Studies of Culture.

Sabatier, P. and Jenkins-Smith, H.C. (eds) (1993a) *Policy Change and Learning: An Advocacy Coalition Approach.* Boulder, CO: Westview Press.

Sabatier, P. and Jenkins-Smith, H.C. (1993b) 'The study of public policy processes', pp. 1–9 in P. Sabatier and H.C Jenkins-Smith (eds), *Policy Change and Learning:. An Advocacy Coalition Approach.* Boulder, CO: Westview Press.

Sahlin, K. and Wedlin, L. (2008) 'Circulating ideas: imitating, translation and editing', pp. 218–42 in R. Greenwood, C. Oliver, K. Sahlin and R. Suddaby (eds), *The SAGE Handbook of Organizational Institutionalism.* Thousand Oaks, CA: Sage.

Sahlin-Anderson, K. and Engwall, L. (2002) 'Carriers, flows and sources of management knowledge', pp. 3–32 in K. Sahlin-Andersson and L. Engwall (eds), *The Expansion of Management Knowledge.* Stanford, CA: Stanford Business Books.

Salamon, L.M (2002) 'The New Governance and the tools of public action: an introduction', pp. 1–47 in L.M Salamon (ed.), *The Tools of Government: A Guide to the New Governance.* Oxford: Oxford University Press.

Sartori, G. (2009) 'Concept misformation in comparative politics', pp. 13–43 in D. Collier and J. Gerring (eds), *Concepts and Method in Social Science: the Tradition of Giovanni Sartori.* London: Routledge [first printed in *American Political Science Review* 64(4) 1970:1033–53].

Scharpf, F. (1977) 'Does organization matter? Task structure and interaction in the ministerial bureaucracy', pp. 149–67 in E. Burack and A. Negandhi (eds), *Organization Design.* Kent, OH: Kent State University Press.

Schattschneider, E.E. (1960) *The Semisovereign People.* New York: Holt, Rinehart and Winston.

Schimmelfennig, F. and Rittberger, B. (2006) 'Theories of European integration: assumptions and hypothesis', pp. 73–95 in J. Richardson (ed.), *European Union: Power and Policy-making*. Abingdon: Routledge.

Schmitter, P.C. (1974) 'Still the Century of Corporatism?' *Review of Politics*, 36(1): 85–131.

Schneiberg, M. and Bartley, T. (2001) 'Regulating American industries: markets, politics, and the institutional determinants of fire insurance regulation', *American Journal of Sociology*, 107(1):101–46.

Schneiberg, M. and Clemens, E.S. (2006) 'The typical tools for the job: research strategies in institutional analysis', *Sociological Theory*, 24(3):195–221.

Schneider, M. and Teske, P. (1992) 'Toward a theory of the political entrepreneur: evidence from local government', *American Political Science Review*, 86(3):737–47.

Schreurs, M.A. (2002) *Environmental Politics in Japan, Germany, and the United States*. Cambridge: Cambridge University Press.

Scott, W.R. ([1995] 2008) *Institutions and Organizations: Ideas and Interests*, 3rd edn. Thousand Oaks, CA: Sage.

Scott, W.R., Ruef, M., Mendel, P.J. and Caronna, C.A. (2000) *Institutional Change and Healthcare Organizations*. Chicago, IL: University of Chicago Press.

Selin, H., and VanDeveer, S. (2009) *Changing Climates in North American Politics: Institutions, Policymaking, and Multilevel Governance*. Cambridge, MA: MIT Press.

Selznick, P. (1957) *Leadership in Administration*. New York: Harper and Row.

Semerklæringen (2001) *Det politiske grunnlaget til Bondevik II–regjeringen, 8. oktober 2001*. Oslo: KrF (Christian Democratic Party).

SFT (1999) Utslippstillatelse for Naturkraft AS-gasskraftverk på Kollsnes 21. januar 1999. Available at www.sft.no/nyheter/dokumenter/gasskraft/kollsnes-utslippstillatelse.html (Accessed 1 October 2004). Oslo: Statens forurensningstilsyn (now: Norwegian Climate and Pollution Agency).

SFT (2006) *Etablering av gasskraftverk på Mongstad*. Statens forurensingstilsyns anbefaling til Miljøverndepartementet. Oslo: Statens forurensningstilsyn (now: Norwegian Climate and Pollution Agency).

Shell (2009) *Delivery and Growth Report*. Royal Dutch Shell PLC Annual Report and Form 20 F for the Year Ended 31 December, 2008. The Hague and London: Shell.

Simon, H. (1997 [1947]) *Administrative Behavior: A Study of Decision-making Processes in Administrative Organizations*. New York: Free Press.

Sims, R.E.H, Schock, R.N., Adegbululgbe, A., Fenhann, J,. Konstantinaviciute, I. et al. (2007) 'Energy supply', Chapter 4 in B. Metz, O.R. Davidson, P.R. Bosch, R. Dave and L.A. Meyer (eds), *Climate Change 2007: Mitigation*. Cambridge: Cambridge University Press.

Skanska (2008) Skanskas historie. Available at: www.skanska.no/no/Om-Skanska/Skanskas-historie/. (Accessed 20 August 2008).

Skocpol, T. (1985) 'Bringing the state back in: strategies of analysis in current research', pp. 3–28 in P.B Evans, D. Rueschemeyer and T. Skocpol (eds), *Bringing the State Back In*. Cambridge: Cambridge University Press.

Skocpol, T. (2003) 'Doubly engaged social science', pp. 407–28 in J. Mahoney and D. Rueschemeyer (eds), *Comparative Historical Analysis in the Social Sciences*. Cambridge: Cambridge University Press.

Skowronek, S. (1993) *The Politics Presidents Make: Leadership from John Adams to George Bush*. Cambridge, MA: Belknap Press of Harvard University Press.

Smelser, N.J. (1973) 'The methodology of comparative analysis', pp. 42–86 in D.P. Warwick and S. Osherson (eds), *Comparative Research Methods*. Englewood Cliffs, NJ: Prentice-Hall.

Snow, D. and R. Benford (1988) 'Ideology, frame resonance, and participant mobilization', pp. 197–218 in B. Klandermans, H. Kriesi, and S. Tarrow (eds), *From*

Structure to Action: Comparing Social Movement Research Across Cultures (International Social Movement Research, Volume 1). Greenwich, CT: JAI Press.

Solomon, S., Qin, D., Manning, M., Chen, Z. Marquis, M., et al. (eds) (2007) *Climate Change 2007—The Physical Science Basis.* Working Group I Contribution to the Fourth Assessment Report of the IPCC Intergovernmental Panel on Climate Change. New York: Cambridge University Press.

Solør Bioenergi (2009) *Om Solør bioenergi.* Available at: www.solorbioenergi.no/ (Accessed 7 April 2009).

Somanthan, E., Sterner, T. and Sugiyama, T. (2014) 'National and sub-national policies and institutions', Chapter 15 in Intergovernmental Panel on Climate Change Working Group III Mitigation of Climate Change. Cambridge: Cambridge University

Soria Moria (2005) *Politisk plattform for en flertallsregjering utgått av Arbeiderpartiet, Sosialistisk Venstreparti og Senterpartiet.* [Also called 'Soria Moria Declaration'.] Available at: www.regjeringen.no/nb/dep/smk/dok/rapporter_planer/ rapporter/2005/ soria-moria-erklaringen.html?id=438515 (Accessed 31 January 2011).

SSB (Statistics Norway) (2009) Bygge- og anleggsvirksomhet, strukturstatistikk. Available at: www.ssb.no/stbygganl/tab-2009–05–06–02.html (Accessed 20 September 2009).

Statkraft (2001) *Årsrapport 2000.* Lysaker, Norway: Statkraft.

Statkraft (2005) *Årsrapport 2004.* Lysaker, Norway: Statkraft.

Statkraft (2007) *Årsrapport 2006.* Lysaker, Norway: Statkraft.

StatoilHydro (2009) *Masterplan for Mongstad.* Oslo: StatoilHydro

Statoil (2010) *Årsrapport 2009.* Stavanger: Statoil.

Stiglitz, J.E and Walsh, C.E. (2006) *Economics,* 4th edn. New York: W.W. Norton.

Stinchcombe, A.L. (2001) *When Formality Works.* Chicago, IL: University of Chicago Press.

Stocker, T.F, Quin, D., Plattner, G., Tignor, M.M.B, Allen, S. K., et al. (2014). *Climate Change 2013. The Physical Science Basis.* Working Group I Contribution to the Fifth Assessment Report of the Intergovernmental Panel on Climate Change. Cambridge: Cambridge University Press.

St. meld. nr. 29 (1997–1998) *Norges oppfølging av Kyotoprotokollen.* Oslo: Ministry of the Environment.

St. meld nr. 29 (1998–1999) *Om energipolitikken.* Oslo: Ministry of Petroleum and Energy.

St. meld nr. 37 (2000–2001) *Om vasskrafta og kraftbalansen.* Oslo: Ministry of Petroleum and Energy.

St. meld. nr. 15 (2001–2002) *Tilleggsmelding til St. meld. nr. 54 (2001–2002) Norsk klimapolitikk.* Oslo: Ministry of the Environment.

St. meld. nr. 54 (2001–2002) *Norsk klimapolitkk.* Oslo: Ministry of the Environment.

St. meld. nr. 9 (2002–2003) *Om innenlands bruk av naturgass.* Oslo: Ministry of Petroleum and Energy.

St. meld. nr. 23 (2003–2004) *Om boligpolitikken.* Oslo: Ministry of Local Government and Regional Development.

St. meld. nr. 11 (2006–2007) *Om støtteordningen for elektrisitetsproduksjon fra fornybare energikilder.* Oslo: Ministry of Petroleum and Energy.

St. meld. nr 34 (2006–2007) *Norsk klimapolitikk.* Oslo: Ministry of the Environment.

Stoltenberg, Jens (2007) Statsministerens nyttårstale, 1 January 2007. Available at: www.regjeringen.no/nb/dep/smk/aktuelt/taler_og_artikler/statsministeren/ statsminister_jens_stoltenberg/2007–4/statsministerens-nyttarstale-2007.html? id=440349 (Accessed 5 May 2010).

Stortinget (1997) *Referat fra Stortingets møte, onsdag 4. juni kl. 10 1997.* Oslo: Stortinget.

St. prp nr. 79 (2003–2004) *Om samtykke av godkjenning av avgjerd i EØS-komiteen nr. 37/2004 av 23. april 2004 om innlemming av EØS-avtale av direktiv 2002/91/ EF om energieffektivitet i bygningar.* Oslo: Ministry of Foreign Affairs.

St. prp. nr. 49 (2006–2007) *Samarbeid om håndtering av CO₂ på Mongstad.* Oslo: Ministry of Petroleum and Energy.

St. prp. nr. 38 (2008–2009) *Investering i teknologisenter for CO₂-håndtering på Mongstad.* Oslo: Ministry of Petroleum and Energy.

St. prp. nr. 67 (2008–2009) *Tilleggsbevilgninger og omprioriteringer i statsbudsjettet 2009.* Oslo: Ministry of Finance.

Støa, E., Kittang, D. and Andersen, I. (2006) *Verktøy for miljøprogrammering for tiltak i byer og tettsteder.* Report. Trondheim: SINTEF.

Stokke, O. S (2012) *Disaggregating International Regimes: A New Approach to Evaluation and Comparison.* Cambridge, MA: MIT Press.

Streeck, W. and Thelen, K. (2005) *Beyond Continuity: Institutional Change in Advanced Economies.* Oxford: Oxford University Press [Kindle version].

Thelen, K. (2003) 'How institutions evolve: insights from comparative historical analysis', pp. 208–40 in J. Mahoney and D. Rueschemeyer (eds), *Comparative Historical Analysis in the Social Sciences.* Cambridge: Cambridge University Press.

Thornton, P.H. (2004) *Markets from Culture. Institutional Logics and Organizational Decisions in Higher Education Publishing.* Stanford, CA: Stanford University Press.

Thornton, P.H. and Ocasio, W. (1999) 'Institutional logics and the historical contingency of power in organizations', *American Journal of Sociology,* 105(3):801–43.

Thornton, P.H. and Ocasio, W. (2008) 'Institutional logics', pp. 99–129 in R. Greenwood, C. Oliver, K. Sahlin and R. Suddaby (eds), *The SAGE Handbook of Organizational Institutionalism.* Thousand Oaks, CA: Sage.

Thornton, P.H., Ocasio, W. and Lounsbury, M. (2012) *The Institutional Logics Perspective: A New Approach to Culture, Structure and Process.* Oxford: Oxford University Press.

Thue, L. (1996) *Strøm og styring.* Oslo: Ad NotamGyldendal.

Thyholt, M. (2006) *Varmeforsyning til lavenergiboliger i områder med fjernvarmekonsesjon.* PhD thesis. Trondheim: Norwegian University of Science and Technology (NTNU).

Tjernshaugen, A. (2007) *Gasskraft: tjue års klimakamp.* Oslo: Pax.

Torget, L.M. (2004) *Å lede andre.* Thesis in sociology. Oslo: University of Oslo, Department of Sociology and Human Geography.

Total (2009) *Carbon Dioxide Capture and Geological Storage.* www.total.com/en/ challenges/carbon-dioxide-capture-and-geological-storage/total-s-commitment/ research-programs-and-industrial-projects-940766.html (Accessed 5 February 2010).

Townshend, T., Fankhauser, S., Ayabar, R., Collins, M. Landesman, T., Nachmany, M. and Pavese, C. (2013) 'How national legislation can help to solve climate change, *Nature of Climate Change,* 3:430–32.

Underdal, A. (2002) 'One question, two answers', pp. 3–45 in E.L. Miles, A. Underdal, S. Andresen, J. Wettestad, J.B. Skjærseth and E.M. Carlin (eds) *Environmental Regime Effectiveness.* Cambridge, MA: MIT Press.

Vedder, H. 2003. *Competition Law and Environmental Protection in Europe: Towards Sustainability?* Groeningen: Europa Law Publishing.

Visier, J.C., Thomsen, K.E. and Johanssen, G. (2003) *Energy Performance of Buildings.* ENPER-TEBUC report. St-Stevens-Woluwe: Belgian Building Research Institute.

Watanabe, R. (2011) *Climate Policy Changes in Germany and Japan: A Path to Paradigmatic Policy Change.* London: Routledge.

Weale, A., Pridja, G., Williams, A. and Porter, M. (1996) 'Environmental adminis-tration in six European states: secular convergence or national distinctiveness'? *Public Administration,* 74:255–74.

White, H.C. (1981) 'Where do markets come from?', *American Journal of Sociology,* 87(3):517–47.

Wiener, A. and Diez, T. ([2004] 2009) *European Integration Theory.* Oxford: Oxford University Press.

Wilson, J. Q. (1980) *The Politics of Regulation.* New York: Basic Books.

Wittneben, B.B.F., Chukwumerije, O. Okereke, S., Banerjee B. and Levy, D. L. (2012) 'Climate change and the emergence of new organizational landscapes', *Organiza-tion Studies,* 33(11):1431–50.

Wooten, M. and Hoffman, A. J. (2008) 'Organizational fields: past, present and future', pp. 130–47 in R. Greenwood, C. Oliver, K. Sahlin and R. Suddaby (eds), *The SAGE Handbook of Organizational Institutionalism.* Thousand Oaks, CA: Sage.

Wurzel, R. and Connelly, J. (eds) (2011) *The European Union as a Leader in Inter-national Climate Change Politics.* London: Routledge.

Yin, R.K. (1994) *Case Study Research: Design and Methods.* 2nd edn. London: Sage.

Index

Page numbers in italic format indicate figures and tables.

1000 Flowers mechanism 55, 57–8, 84, 89, *124*, 177

Agder Energi 120, 122
agenda setting skills 14, 66, 67, 70, 75
Aker Clean Carbon 94, 100
Akershus Energi 122
authority distribution 32–3, 40, 49
authority structures 32, 48, *55*, 66, 191

Baumgartner, Frank 26, 37
Bellona foundation 86, 90, 93, 100, 101, 105
biomass 111, 117, 118, 131
bottom-up harmonization *55*, 58, *124*, 171, 172
Building Agency 41, 140, 150, 151
building code 135–40, 143–5, 147–8, 154–5
building construction 134–5, 137–40, 144–5, 148–9
business power 186–90

cap on emissions 8, 9
carbon capture and storage (CCS) policy: assessment of 99–107; conclusion about 107–8; cost issues 97; entrepreneurship and 194–5; European environment and 106–7, 171; introduction to 83–4; mechanisms in *99*; organizational fields and 103–6, 167–8, 187–9; outcomes in 84–5, *162*; political competition and 92–9, 166–7; political consensus 85–9; political fields and 99–103, 164–5, 173–5,

180–1; as a political objective 89–92; promotion of 17; turning point in 94
carbon emissions 35, 86, 89, 137, 140
carbon mitigation 8, 105, 202, 203, 204
case-study techniques 16–20
causal processes 18, *19*, *20*, 196
Centre Party 86, 91, 93, 119
Christian Democrats 86, 87, 91, 92, 101
Christoff, Peter 5, 6, 7
civil servants 77, 154–5, 171–2, 184–7, 189–91
climate change 2, 28–9
climate mitigation 6, 50, 204
climate policy: authority distribution and 32–3; barriers to 66; capturing variation in 7–11; compromise for 96; development of 29; emergence and change of 2–3; energy policy for buildings and 137–48; entrepreneurship and 3, 62; European environment and 12–13; explanatory factors for 5; funding issues and 31–2; information distribution for 32–3; mechanism variations in 36–7; new agenda for 196–7; organizational fields and 46, 201; outcomes in *162*; policymakers and 60–1; political fields and 38, 201; political logics related to *41*; professional logics 50–2; segmentation approaches and 48; societal costs

minimization logic and 51–2, 88;
state of the art and 3–6; turning
point in 94; typology of 8, 10;
see also multi-field approach
CO₂ emissions 85, 88, 92
cognitive frameworks 14, 35, 68, 70, 78
Cohen, Michael D. 43, 44
collaboration mechanism 53, *124, 167*
comparative research 4, 7, 16
Conservative/Liberal government 141,
142
Culpepper, Pepper 31, 41

decision-making 26, 40, 44, 74–5
direct state steering 7, *8*, 135, *162*
distribution of authority 32–3, 40, 49
distribution of information 32–3, 40, 49
distribution of power 39, 53
distribution of structural resources
40–1, 184
district heating policy: development
of 109–10, 113, 131; Enova
scheme and 130; sources for
111; success for 112–19; support
for 122–3
Dutch green certificates *see* green
certificates schemes
dwelling production 144

Eckersley, Robin 5, 6, 7
EFTA Surveillance Agency (ESA) 19,
96–7, 117–18, 146–8, 153
Eidsiva 122
election issues 38–40, 42, 60, 86, 150,
153
electricity price issues 118, 119
emissions permit 86, 87, 93, 101
emissions trading 9, *51*, 57, 83
Emissions Trading System (ETS) 51,
92, 191
energy certification scheme 136–7,
141–2, 147, 153, 155
Energy Law 87, 98
energy performance of buildings *see*
energy policy for buildings
Energy Performance of Buildings
Directive (EPBD) 140, 142, 143,
147
energy policy for buildings: assessment
of 148–55; business power
and 189; climate policy and
137–48; conclusion about
155–6; entrepreneurship and
175; European environment

and 172; introduction to 17,
134–5; mechanisms in *149*;
organizational fields and 169–70,
187; policy outcomes 135–7,
163; political fields and 149–51,
165
Enhanced Oil Recovery (EOR) 93
Enova scheme: energy policy for
buildings and 136, 141–2, 145–8;
renewable energy policy and
111, 112, 123, 126, 130
entrepreneurship: *ad hoc* 153; climate
policy and 3, 62; conceptualizing
63–71; conclusion about 78–9;
defined 70; energy policy for
buildings and 175; European
environment and 170–3, 177–9;
Europeanization and 190–3,
205; Fashion Queen 73–4, 77,
102, 192; green certificates
schemes and 128–30, 169, 172;
Importer 71–3, 101, 128, 152;
institutional 64, 68–9, 105,
129; knowledge status 63–6;
mechanisms 71–8; multi-field
13–15, 65–6, 69–71, 194;
paradoxes 193–6; pluralism and
176; political 100–1, 174–5;
political competition and 173–5;
renewable energy policy and
132–3, 195; Shrewd Lawyer
74–6, 102, 130, 153, 191,
192; Spider 76–7, 101, 128,
152–3; structural 66–7, 78, 129;
understanding the effect of 65
environmental groups 66, 91, 95, 128
environmental issues 28–9
environmental movement 100, 101,
103, 108
environmental policies 7, 9, 38, 53, 88,
183
ETS Directive 92
EU Governing *55*, 57, 129, 179
EU Renewable Energy Directive 116,
122
European Committee for
Standardization (CEN) 145
European Economic Area (EEA) 13, 17,
18, 102, 172
European Energy Performance of
Buildings Directive (EPBD) 140,
141, 143, 147
European environment: CCS policy
and 106–7, 171; climate policy

and 12–13; comparison with other fields 38–60, 164–73; description of 54; energy policy for buildings and 154–5, 172; power distribution and 53; renewable energy policy and 131–2, 171–2; social mechanisms in 53–8; structural dimensions of 30–3, 55; *see also* entrepreneurship
'European integration theory' 54, 55
Europeanization: defined 55; institutional features 56; literature on 54; policymaking/policymakers and 53, 55; structural effects of 56; *see also* entrepreneurship
European utilities 115, 131, 172, 181

Fashion Queen entrepreneurship *see* entrepreneurship
feed-in support schemes 10, 115, 116, 118, 121, 131
fields: description of 27, 60; institutional dimensions of 34–5, 41, 49–50; range of 28; social architecture of 26–37; structural dimensions of 30–3, 39, 48–9; *see also* organizational fields; political fields
fiscal incentives 8, 10, 110–11, 135, 136
Fligstein, Neil 26, 27, 65
funding issues 31–2

garbage can processes: beware of 204–5; features of 43–4; organizational fields and 189; political fields and 166; politicians and 44–6, 125, 150, 183, 186
gas-power plants 84–6, 90–3, 104, 137
gas prices, rising 97, 118
Gassnova 84, 91, 94, 105, 108
'goodness-of-fit' explanations 71, 73, 191
Governmental Industry Development approach 8–10, 84–5, 102, 107, 135
green certificates schemes: climate policy and 51; demand for 116; electricity prices and 119; entrepreneurship and 128–30, 169, 172; lobby pressures for 114; market measures and 9; for

renewable energy policy 111, 113, 125, 163; Sweden and 121, 122
greenhouse gas (GHG) emissions 6, 9

Hafslund 122
Haga, Åslaug 100, 102
high energy-performance buildings *see* energy policy for buildings
holistic approach 136, 139, 141, 142
Housing Bank 135, 138–40, 149, 151–2, 155
Hydro (company) 86, 103, 168, 180
hydropower plants 109, 114
Hydro project 85, 86, 103, 180

Importer entrepreneurship *see* entrepreneurship
indirect state steering 8, *162*
information alteration 76
information distribution 32–3, 40, 49
institutional dimensions of fields 34–5, 41
institutional entrepreneurship *see* entrepreneurship
institutional logics: CCS policy and 104; cognitive elements and 35; entrepreneurship and 68, 69, 75, 78; institutional elements and 34; multiple 36, 50
institutional spheres *see* fields
instructional logics 199, 202–3
interest groups 46, 47, 72, 187
International Panel on Climate Change (IPCC) 4, 5, 6, 7, 89
international policies 4, 12
iron triangles 12, 47, 103, 167, 187

Jones, Bryan 26, 37

Kårstø gas processing plant 84–6, 92–3, 97–8, 104
Kingdon, John 70, 184
Kyoto Protocol 53

Labour Party 85, 86, 90, 91, 100
land-use planning regulations 9
layering, description of 163
Legislature Governing 39, 44–5, 103, 164–5
Lindeberg, Erik 86, 91
lobbying activities/efforts 66, 91, 113, 115, 120
logic of minimizing societal costs *see* societal costs minimization logic

low-carbon investments 51, 163
low-carbon technologies 96, 202
low-energy building concept 134, 135, 138, 148, 152

March, Jim 15, 35, 43
market logic: building construction and 138–9, 150–1; CCS policy and 103–4; climate policy 50–1; petroleum corporations and 87
market measures: description of 8–9; renewable energy policy 110–11; *see also* green certificates schemes
McAdam, Doug 26, 27
minimizing societal costs logic *see* societal costs minimization logic
Ministerial Governing 39, 45, *124*, 125, 166
Ministry of Local Government and Regional Development 135, 142, 143, 144, 148, 150
Ministry of Local Government and Regional Planning 137, 140, 152
Ministry of Petroleum and Energy (MPE) 86, 91, 117, 129, 153, 155
Ministry of the Environment 86–7, 89–90, 125, 137
Minogue, Martin 38, 40
misfit situations 71
Mongstad agreement 94
Mongstad refinery 84, 93
'moon landing' metaphor 95, 98, 102
multi-field approach: analytical value of 26; business power and 186–90; case studies techniques and 16–18; climate policy and 3–15, 196–7; conclusion about 20–1; different aspects of 29–30; entrepreneurship and 13–15, 37, 65–6, 69–71, 194; future research for 198; introduction to 1–3; pattern matching and 19–20; pervasive trends and 1–2; policymaking process and 11–13; power of politics and 182–6; process tracing and 18–19; theory blending and 15–16
multi-level governance approach 14, 15, 26, 54, 56
municipally-owned companies 113, 114

Naturkraft 85, 86, 87, 91, 92, 97
negative framing 69, 72, 73

neo-institutional literatures 28, 52, 57, 62, 193
neo-pluralists 47, 48, 127, 187
networks/networking: entrepreneurship through 47, 67, 75; pluralism and 47; policy 55, 56
norms, purpose of 34–5
Norwegian cases: CCS policy 83–108; comparative assessment of 161–81; energy policy for buildings 134–56; renewable energy policy 109–33; selection of 17–18
Norwegian Pollution Authority 87, 162
Norwegian-Swedish scheme 120, 126
Norwegian utilities 109, 112, 181

offshore activities 87, 98
oil-burning furnaces 136
oil corporations 90, 94
Olsen, Johan P. 15, 26, 35, 43
organizational fields: actors representing 36; business power and 186–90; CCS policy and 103–6, 167–8, 187–9; climate policy and 46, 201; comparison with other fields 58–60, 167–70; energy policy for buildings and 151–4, 169–70; entrepreneurs and 175–7; importance of 170; information distribution and 49; institutional change and 50; mapping of 200–1; multi-field approach and 12; petroleum industry and 84; political fields and 179–81; renewable electricity and 127–32, 169; renewable heating and 168–9; social mechanisms in 46–53, 60
organizations 28, 31–3, 51–2, 104
Organization Studies 28

Palm, Thomas 91, 92
passive houses 134, 136, 139, 140, 145
pattern matching 18–20
Peaceful Collaboration situations *175*, 176
Petroleum Directorate 93, 98
petroleum industry 84, 86–7, 94, 98, 203
Petroleum Law 87, 98
pluralism: case-study techniques and 16; entrepreneurship and 176; organizational fields and 187–8; policymaking and 46–8, 52–3; theoretical 197

policymaking/policymakers: advice to 199–205; approaches to 47; business power and 186–90; climate policy and 60–1; Europeanization and 53, 55; as a matter of persuasion 68; organizational fields and 200–1; pluralist view on 46–8, 52–3; political fields and 38; structural resources distribution and 40–1, 184; *see also* entrepreneurship

policy monopolies 1, 12, 47, 103, 167, 187

political competition 41–3, 92–9, 166–7, 173–5

political entrepreneurship *see* entrepreneurship

political fields: actors representing 36; CCS policy and 99–103, 108, 164–5, 173–5, 180–1; climate policy and 38, 201; comparison with other fields 58–60, 164–7; energy policy for buildings and 149–51, 165; mapping the position of issue in 201–2; multi-field approach and 11–12; organizational fields and 179–81; petroleum industry and 84; renewable energy policy and 125–7, 180; social mechanisms in 38–46, 60

politicians: climate policy and 41; garbage can processes and 43–6, 125, 150, 183; importance of 185; Legislature Governing and 44–5; main challenge for 42; Ministerial Governing and 45; politicizing and 45; Random Steering and 45, 46

Politicizing situations 39, 99–100, 124–5, 174

politics, power of 182–6

Pollution Authority 87, 89

Pollution Control Act 92, 101, 162

pollution permits 8, 57, 88

positive framing 68–9, 72, 73

power distribution 39, 53

power producers 114–15, 118, 127, 130

process tracing 18–20, 99, 152

professional logics, types of 50–2

Progress Party 95, 100, 145

Quiet Politics and Business Power (Culpepper) 188

R&D issues 5, 86, 91, 98, 105, 108

Random Steering 45, 46

red/green coalition 96, 119, 121, 143, 153

renewable electricity: case selection 17; entrepreneurship and 132–3, 195; mechanisms in *124*; organizational fields and 127–32, 169; policy outcomes 162–3; political fields and 165–6

renewable energy policy: assessment of 124–31; climate policy and 10; conclusion about 132–3; electricity prices and 119–23; entrepreneurship and 132–3; European environment and 131–2; introduction to 109–10; as a main priority 123–4; policy outcomes 110–11; political attention to 118–19; political fields and 125–7, 180; puzzling nature of 110; societal costs minimization logic and 114; success for 112–23

renewable heating: case selection 17; Enova scheme for 112; European environment and 171; low salience 132; mechanisms in *124*; organizational fields and 168–9; policy outcomes 163; political fields and 126, 165; state aid for 110–12, 116–17

Scott, Dick 34, 35

Second World War 46, 134, 138

segmentation approaches 47, 48, 175–7, 187–8

Shrewd Lawyer entrepreneurship *see* entrepreneurship

SINTEF Petroleum Research 86, 91

Skocpol, Theda 15, 37

Socialist Left Party 86, 93

social mechanisms: comparing 58–60; conclusion about 60–1; entrepreneurship and 193–6; in European environment 38–60; features of 36–7; importance of 13–14; interrelationships between 173–81; introduction to 25; in organizational fields 46–53, 60; in political fields 38–46; social architecture of fields and 26–37

societal costs minimization logic: climate policy and 51–2, 88;

energy policy for buildings and 146, 150, 151; gas-power opponents and advocates and 90; renewable energy policy and 114, 120; support for 95
societal fields 1, 31, 36, 39
Spider entrepreneurship *see* entrepreneurship
state aid: CCS policy 96, 97, 102, 106–7; for renewable heating 110–12, 116–17; for wind power 109
State Housing Bank 135, 138–40, 149, 151–2, 155
state-owned utilities 115
state steering 7–8, 161–2
stationary energy production 108, 109, 114, 168, 169
Statkraft 113, 115, 127–30, 168–9
Statoil 85–7, 90, 93, 98, 103–5, 180
StatoilHydro 85, 98
Steensnæs, Einar 92, 100, 101
stock exchange 130, 138, 168
Stoltenberg, Prime Minister Jens 90, 95, 96, 101–2, 120
structural entrepreneurship *see* entrepreneurship
structural patterns 39, 48, 55
structural resources distribution 40–1, 184
Sweden: green certificates schemes and 113, 116, 121, 122; negotiations with 119, 129, 181

technological innovation 52
Technology Centre Mongstad, TCM 84, 85, 96, 97, 98

technology development logic: aligning toward 96, 97, 98; building construction and 138, 140, 145, 150, 151; gas-power opponents and advocates and 90–1; petroleum corporations and 87; renewable energy policy and 114, 118, 122
technology standards 8–9, 52, 111, 135, 139
theory blending 15–16
theory of fields 11, 26, 29
theory pluralism 16, 197
thermal qualities of buildings 135, 139–41
Turf Battle mechanism 53, *124*, *167*, 176
typology: of climate policy 8, 10; purpose of 10

unpredictable EU Governing 55, *124*, *149*, 178–9
utilities, large 123, 130

Viken district heating 123
voluntary agreements 9

'Washington establishment' 26
waste heat 111
wind power conflicts 112–19
wind power development 109–10, 120
wind power projects 116, 119

'zero-energy buildings' 134
ZERO foundation 93, 105